PRAISE FOR **PROJECTIONS**

"Because of his experiences as a physician and researcher, Dr. Deisseroth recognizes the limitations of science and medicine and the transcendent value of elemental human connection. . . . In life's most difficult moments, it might be everything."

—*The Wall Street Journal*

"[A] scintillating and moving analysis of the human brain and emotions . . . A great read."

—*Nature*

"[Karl Deisseroth's] imaginative narrative flows effortlessly. . . . There is a first love of reading and writing and hints of a literary imagination that draws on James Joyce and Toni Morrison. . . . His narratives are always sensitive. . . . An admixture of fact and fiction, reality and imagination, damage and desire."

—*Science*

"*Projections* is not only a joy to read; it is also a compelling, educational experience."

—*Psychiatric Times*

"*Projections* asks probing questions about some of our most fundamental human traits to shed light on the origins of our emotions. Why, for instance, do we shed tears? How did this show of weakness survive millennia of evolution? Deisseroth writes of heartbreaking and desperate medical cases with a doctor's knowledge and a novelist's skill for narrative. I was fascinated, and could not put this book down."

—MAY-BRITT MOSER, Nobel Laureate

"I find myself at a loss for how to describe this remarkable work. Just as Karl, through his laboratory, has reimagined, and literally redefined how we view the human brain, he has reimagined and redefined what literary nonfiction can be, with great elegance. For all of us who write about science for the public, this will be a tough act to follow. It's poetic, mind-stretching, and through it all, deeply human. I am driven to the brink of despair: whether to chuck my writing career altogether, or double down and work like a madman to try and write something half this good."

—DAN LEVITIN, *New York Times* bestselling
author of *The Organized Mind*

"We are living during a revolution in our understanding of the human brain, and Karl Deisseroth has been at the forefront of these advances. This magisterial work, *Projections*, shows that not only is he one of our leading scientists, but also a gifted writer and storyteller. With precise yet luminous prose, he merges stories of cutting-edge neuroscience with a deep reverence for his patients' humanity."

—NEIL SHUBIN, author of *Some Assembly Required*

"Unique and utterly riveting . . . This is a masterpiece written for each and every one of us."

—PATRICIA CHURCHLAND, author of *Conscience*

PROJECTIONS

PROJECTIONS

The New Science
of Human Emotion

KARL DEISSEROTH

RANDOM HOUSE
NEW YORK

2023 Random House Trade Paperback Edition

Copyright © 2021 by Karl Deisseroth

All rights reserved.

Published in the United States by Random House, an imprint and
division of Penguin Random House LLC, New York.

RANDOM HOUSE and the HOUSE colophon are registered trademarks
of Penguin Random House LLC.

Originally published in hardcover in the United States by
Random House, an imprint and division of Penguin Random House LLC,
in 2021.

Permissions credits can be found on page 233.

LIBRARY OF CONGRESS CATALOGING-IN-PUBLICATION DATA
Names: Deisseroth, Karl, author.
Title: Projections / by Karl Deisseroth.
Description: First edition. | New York: Random House, [2021]
Identifiers: LCCN 2020026946 (print) | LCCN 2020026947 (ebook) | ISBN
9781984853714 (paperback) | ISBN 9781984853707 (ebook) | ISBN
9780593448168 (international edition)
Subjects: LCSH: Client-centered psychotherapy—Case studies. |
Emotions. | Mental illness. | Optogenetics.
Classification: LCC RC481 .D45 2021 (print) | LCC RC481 (ebook) |
DDC 616.89/14—dc23
LC record available at https://lccn.loc.gov/2020026946
LC ebook record available at https://lccn.loc.gov/2020026947

Printed in the United States of America on acid-free paper

randomhousebooks.com

1st Printing

Book design by Simon M. Sullivan

For our family

I offer you the memory of a yellow rose seen at sunset, years before
you were born.
I offer you explanations of yourself, theories about yourself, authentic
and surprising news of yourself.
I can give you my loneliness, my darkness, the hunger of my heart; I
am trying to bribe you with uncertainty, with danger, with defeat.

—JORGE LUIS BORGES, "Two English Poems"

CONTENTS

PROJECTIONS

PROLOGUE

In the art of weaving, warp threads are structural, and strong, and anchored at the origin—creating a frame for crossing fibers as the fabric is woven. Projecting across the advancing edge into free space, warp threads bridge the formed past, to the ragged present, to the yet-featureless future.

The tapestry of the human story has its own warp threads, rooted deep in the gorges of East Africa—connecting the shifting textures of human life over millions of years—spanning pictographs backdropped by crevassed ice, by angulated forestry, by stone and steel, and by glowing rare earths.

The inner workings of the mind give form to these threads—creating a framework within us, upon which the story of each individual can come into being. Personal grain and color arise from the cross-threads of our moments and experiences, the fine weft of life, embedding and obscuring the underlying scaffold with intricate and sometimes lovely detail.

Here are stories of this fabric fraying in those who are ill—in the minds of people for whom the warp is exposed, and raw, and revealing.

•

The bewildering intensity of emergency psychiatry provides a context for all of the stories in this volume. If such a setting is to illuminate the shared fabric of the human mind, disrupted inner states should be rendered onto the page as faithfully as possible. So here, symptom descriptions from patients are unaltered and real, to reflect the essential nature, the true timbre and soul, of these experiences—though to maintain privacy, many other details have been changed.

Likewise, the powerful neuroscience technologies described—which complement psychiatry by providing a distinct way of looking into the brain—are also entirely real, despite their sometimes science-fiction-like and decidedly unsettling qualities. As depicted here, these methods are drawn unchanged from peer-reviewed papers out of laboratories around the world, including my own.

But even medicine and science are inadequate alone to describe the human internal experience, and so some of these stories are told not from the point of view of a doctor or a scientist but from a patient's perspective—sometimes in the first or third person, sometimes with altered states reflected by altered language. Where another person's inner depths—their thoughts or feelings or memories—are depicted in this way, the text reflects neither science nor medicine, but only a reaching out of my own imagination, with care and respect and humility, to create a conversation with voices I have never heard, but only sensed in echoes. The challenge of trying to perceive, and experience, unconventional realities from the patient's perspective is the heart of psychiatry, working through the distortions of both observer and observed. But inevitably the true innermost voices of the departed and the silenced, the suffering and the lost, remain private.

Imagination here is of uncertain value, and none is asserted, but experience has revealed the many limitations of modern neuroscience and psychiatry in isolation. Ideas from literature have long seemed to me just as important for understanding patients—at times providing a window into the brain more informative than any microscope objective. I still value literature as much as science in thinking about the mind, and whenever possible I return to a lifelong love of writing—though for years this love was only a banked ember, covered with science and medicine, like drifts of ash and snow.

Somehow three independent perspectives, psychiatry and imagination and technology, together can frame the conceptual space needed—perhaps because they have little in common.

Along the first dimension is the story of a psychiatrist, told through a progression of clinical experiences, each centered on one or two human beings. Just as when a fabric frays, its hidden structural threads can be revealed (or when a bit of DNA mutates, the original functions of the damaged gene can be inferred), the broken describe the unbroken—and so each story underscores how the hidden inner experiences of healthy human beings, and perhaps of a doctor as well, might be revealed by the even more cryptic and shadowed experiences of psychiatric patients.

Each story also imagines the emerging human inner experience of emotions, in moments of the world today and across millennia in steps along our journey, past obstacles in our path that may not have been traversed without compromise. This second progression begins with stories of simple and ancient circuits for just being alive—the cells for breathing, or for moving with muscles, or for creating the fundamental barrier between self and other. That earliest, most primal boundary, between each of us and the world—called ectoderm, a lonely fragile layer thin as a single cell—gives rise to skin as well as brain, and so it is with this same ancient borderline that contact between human beings is felt in all its forms, physical or psychological—across the spectrum, from healthy to disordered social states.

The stories move among the universal feelings of loss and grief in human relationships, to deep fractures in the basic experience of external reality that come with mania and psychosis, and finally to disruptions encroaching even on the inner self: the lost ability to feel pleasure in our lives as can happen in depression, the lost motivation to nourish ourselves as in disorders of eating, and even the loss of the self itself—with dementia at life's end. Along this second dimension, the emotions of the subjective inner world, we begin and end with imagination—whether in stories of prehistory (feelings leave no fossils; we cannot know what was felt in the past, and so we do not attempt to be evolutionary psychologists) or of the present (since even today we cannot directly observe another human being's inner experience).

But where the measurable effects of feelings are consistent across individuals—as far as we can tell with carefully applied technology—experimental insight into the inner workings of the brain may develop.

Along a third dimension, each story reveals this rapidly emerging scientific understanding, with clues from both healthy and disordered states, supported by experiments and driven by data. Brief references, for background on the science within each story, are included in Notes at the end of the book; some curious readers might wish to wander therein, along various trails of personal interest. Many additional important contributions are referenced from within each of these links (and so the links serve primarily as initial stepping-stones, to support further exploration)—but only citations in open-source format are listed in this volume, to ensure availability for all. This final dimension is thus a scientific axis—drawn to guide the public without scientific training, people who deserve to grasp, and own for themselves, every idea and concept here.

This text, then, is not only about the experiences of a psychiatrist, nor imagining the emergence of human emotions, nor even the latest neurotechnology. Each of these three perspectives acts only as a lens, each focused in a different way on the central mystery of feelings in the mind, each providing a different view of the same scene. It's not simple to fuse these disparate perspectives together into a single image—but it is no more easy to be human, or to become humanity—and the volume can, in the end, achieve a sort of grainy resolution.

Profound respect and gratitude is expressed here to my patients, whose challenges have provided us with this perspective—and to all whose inner suffering, known or unknown, has been inextricably part of the long, dusky, desperate, uncertain, and occasionally lovely tapestry of our shared journey.

•

A word about myself, and my own path, may be helpful so that the distortions of the narrator can be better known; I am—as are we all—more subjective than objective, only a flawed bit of human optics. In early life there was no hint that the particular path I was traveling would lead to psychiatry—or that the journey would also wind through the even less congruent realm of engineering.

My childhood was set in an ever-shifting context, from small towns to big cities, from the East to the West to the middle of the North American continent and back again, following my restless family—my mother and father and two sisters, who like me all seemed to value reading above every other pursuit—as we moved every few years to a new home. I remember reading to my father for hours at a time, day after day, as we drove across the country from Maryland to California; my own free moments were occupied mostly by stories and poems—even while cycling to and from school, with the book of the moment held perilously on the handlebars. Though I read history and biology too, the imaginative uses of language seemed more compelling to me, until I collided with a different kind of idea that had been lying in wait along my path.

Creative writing was my first registered course in college, but that year I unexpectedly learned from talking with my fellow students, and then in classes, how a particular style of approaching life science—building understanding from single cells, even for inquiry into the most complex larger-scale systems—was helping to resolve some of the deepest mysteries in biology. These questions had long seemed to be nearly intractable: how a body could develop from a single cell, or how the intricate memories of immunity could be formed and preserved and awakened in scatterings of single cells drifting along blood vessels, or how the disparate causes of cancer—from genes to toxins to viruses—could be unified in a single cell-based concept, in a way that was useful, that mattered.

These diverse fields were all revolutionized by bringing small-scale elemental understanding to large-scale complex systems. The shared secret for biology, it seemed to me, was reaching down to the level of cells and their molecular principles, while retaining perspective on the whole system, the whole body. The inner feeling evoked in me by the prospect of extending this simple cellular idea to the mysteries of the mind—of awareness, of emotions, of the stirring of feeling by language—was a pure and pressured delight, like Toni Morrison's "rogue anticipation with certainty," that universal human state of restless joy upon suddenly seeing a path forward.

In talking with friends in our shared dormitory (fellow students who were all, inexplicably, theoretical physicists) over meals, I discovered this was a feeling shared by cosmologists probing phenomena playing out over astronomical scales of space and time. They too began by considering the smallest and most elemental forms of matter, together with the fundamental forces guiding interactions over tiny distances. The result was a process both celestial and personal. The feeling was of synthesis and analysis, together.

Crucially for what came later, at around the same time I was exposed to neural networks, a rapidly growing branch of computer science in which actual memory storage needing no guidance or supervision is achieved by simple collections of units, each cell-like and elemental—things existing in code, with only simple abstract properties—but connected to one another virtually, by operation of the program. Neural networks, as the name suggests, were inspired by neurobiology, but these ideas were so powerful that this computational field would later spawn a revolution in machine intelligence called deep learning, which today uses large collections of cell-like elements to reshape virtually every area of human inquiry and information— including, returning the original favor, neurobiology.

Large groups of connected small things, it seems, can achieve almost anything—if connected in the right way.

I began to consider the possibility of understanding something as mysterious as emotion, at the level of cells. What causes powerful feelings in the well or ill person, feelings adaptive or maladaptive? Or more directly, what in fact *are* those feelings, in a physical sense, down to the level of cells and their connections? This struck me as perhaps the most profound mystery in the universe—rivaled only by the question of the universe's origin, its reason for being.

Clearly, the human brain would be important in approaching this challenge, because only human beings can adequately describe their emotions. Neurosurgeons (I thought) had the most physical and privileged access to the human brain; therefore, the logical path for me, the one that provided the most direct approach to helping, healing, and studying the human brain, seemed to be neurosurgery. So throughout graduate school and into my medical training, I steered myself in this direction.

However, toward the final year of medical school, like all medical students, I was required to complete a brief rotation in psychiatry, without which I could not graduate.

Until that point, I had never felt any particular affinity for psychiatry; in fact I had experienced the field as unsettling. Perhaps it was the seeming subjectivity of the diagnostic tools available, or perhaps there was within me some unknown, even deeper issue I had not addressed. Whatever the reason, psychiatry was the last specialty I would have selected. On the other hand, my early experiences with neurosurgery had been invigorating: I loved the operating room, the life-or-death drama juxtaposed with meticulous precision and attention to detail, the focus and intensity and rhythm of suturing set against the high charge of arousal. And so it was stunning to my friends and family, and to myself, when I chose psychiatry instead.

I had been trained to see brains as biological objects—as they indeed are—organs built from cells and fed by blood. But in psychiatric illness, the organ itself is not damaged in a way we can see, as we can visualize a fractured leg or a weakly pumping heart. It is not the brain's blood supply but rather its hidden communication process, its internal voice, that struggles. There is nothing we can measure, except with words—the patient's communications, and our own.

Psychiatry was organized around the deepest mystery in biology, perhaps in the universe, and I could only use words, my first and greatest passion, to crack open a gate leading to the mystery. This conjunction, once realized, reset my path entirely. And it all began, as life-changing disruptions so often do, with a singular experience.

•

On the first day of my psychiatry rotation, I was sitting in the nurses' station flipping through a neuroscience journal when, after a brief commotion outside, a patient—a man in his forties, tall and thin, with a sparse, bedraggled beard—burst in through a door that should have been locked. In arm's reach, standing above me, he fixed his gaze to mine—his eyes wide in fear and rage. My gut clenched as he began to shout at me.

Like any city dweller, I was no stranger to people saying strange things. But this was no street encounter. The patient appeared completely alert, not shrouded in a fog; his experience was stable and crystalline, the hurt was bright in his eyes, the terror was real. In a shaking voice that seemed to be all he had left, with profound bravery, he was meeting the threat.

And his speech—it was creative in its agony, full of phrases used not for traditional meaning but seemingly for their own sake, as communicative effects, with their own grammar and aesthetic, self-contained. He was directly confronting me—though we had not met before, he had an idea that I had violated him—but was doing so with sounds as feelings, with their relationships beyond syntax or idiom. He spoke a novel word that sounded like one in a phrase of Joyce's I had read long ago: it was *telmetale*; this was *Finnegans Wake* on the locked unit, as he told what was deeper than skin or skull, than stem or stone. I sat agape, my brain rewiring as he spoke. He evoked in me science and art together, not in parallel but as the same idea, fused: with both the steady inevitability and the uncontrolled blaze of a sunrise. It was shocking, it was unitary, it mattered, and it brought my intellectual life fully together for the first time.

I later learned that he was suffering from something called schizoaffective disorder, a destructive storm of emotion and broken reality that combines the major symptoms of depression, mania, and psychosis. I also learned that this definition mattered not at all, since the categorization had little impact on treatment beyond simply identifying and treating the symptoms themselves, and there was no underlying explanation. Nobody could give answers to the simplest questions regarding what this disease really was in a physical sense, or why this person was the one suffering, or how such a strange and terrible state had come to be part of the human experience.

•

Being human, we try to find explanations, even when that quest seems hopeless. And for me, after that moment, there was no turning back — and the more I learned, no turning away. I formally chose psychiatry as my clinical specialty later that year. After completion of four more years of training and board certification in psychiatry, I launched a lab in a new bioengineering department — at the same university, in the heart of Silicon Valley, where I had been a medical student. I planned to treat patients while also building tools for studying the brain. Perhaps new questions, at least, could be asked.

As complicated as the human brain seems to be, it is only a clump of cells like every other part of the human body. These are beautiful cells, to be sure, including more than eighty billion neurons specialized for conducting electricity, each shaped like a richly branched bare tree in winter — and each forming tens of thousands of chemical connections called synapses with other cells. Tiny blips of electrical activity continually course through these cells, pulsing along fat-insulated electrical-conduction fibers called axons that together form the white matter of the brain, each pulse lasting only a millisecond and measurable in picoamps of current. This intersection of electricity and chemistry somehow gives rise to everything that the human mind can do, remember, think, and feel — and it is all done with cells, which can be studied and understood and changed.

As was required for the modern ascendance of other fields of biology (like development and immunology and cancer), new methods first had to be enabled for neuroscience that would allow deeper cellular understanding within the intact brain. Before 2005, we had no way of causing precise electrical activity to happen in specific cells within brains. To that point cellular-level electrophysiological neuroscience had been largely limited to observation — listening with electrodes to cells as they fired away during behavior. This was an immensely valuable perspective in its own right, but we could not provide or take away those firing events within specific cells to see how cellular activity patterns might matter for the elements of brain function and behavior: sensation, cognition, and action. One of the earliest technologies developed in my laboratory beginning in 2004 (called optogenetics) began to address this limitation: the challenge of causing or suppressing precise activity in specific cells.

Optogenetics begins by transporting foreign cargo—a special kind of gene—as far as imaginable in biology: from the cells of one major kingdom of life all the way to the cells of another. The gene is just a piece of DNA that directs its cell to produce a protein (a small biomolecule designed for certain jobs in the cell). In optogenetics we borrow genes from diverse microbes such as bacteria and single-cell algae, and deliver this alien cargo to specific brain cells of our fellow vertebrates like mice and fish. It's a strange thing to do—but with a certain logic, for the particular genes we borrow (called microbial opsins), upon delivery to a neuron, immediately direct the creation of remarkable proteins that can turn light into electrical current.

Normally these proteins are used by their original microbial hosts to convert sunlight into electrical information or electrical energy—by guiding movement of the free-swimming algal cell to the optimal level of light for survival, or (in certain ancient forms of bacteria) by setting up conditions for harvesting energy from the light. In contrast, most animal neurons normally do not respond to light—there would be no reason to, as it is quite dark inside the skull. With our optogenetic approach (using genetic tricks to produce these exotic microbial proteins *only* in specific subsets of neurons in the brain, but not others), those brain cells newly endowed with microbial proteins become much different from their neighbors. At this point the modified neurons are the only cells in the brain able to respond to a pulse of light delivered by a scientist—and the result is called optogenetics.

Because electricity is a fundamental currency of information in the nervous system, when we send in laser light (delivered through thin fiberoptics, or with holographic displays that project spots of light into the brain) and thereby change the electrical signals flowing through these modified cells, remarkably specific effects on animal behavior result. Discovered in this way are the targeted cells' capabilities for giving rise to mysteries of brain function, like perception and memory. These optogenetic experiments have proved so useful in neuroscience because they allow us to link the local activity of individual cells to the global perspective of the brain. Tests of cause and effect now play out in the right context; only cells within intact brains can give rise to the complex functions (and dysfunctions) underlying behavior—just as individual words only matter for communication within the context of their sentence.

We do this mostly in mice, rats, and fish: animals with many nervous system structures in common with us (structures that are just quite a bit scaled up in our lineage). Like us, these fellow vertebrates sense, and decide, and remember, and act—and in so doing, if observed in the right way, they reveal the inner workings of brain structures that we share. And so a new approach to investigating the brain has emerged, with methods that recruit tiny and ancient achievements of evolution to work for us—borrowed from forms of life that diverged from our own lineage almost at the very beginning, at the earliest and deepest anchor of the warp of life itself.

A subsequent technology my team developed, also inspired by this principle of cellular resolution in intact brains, is called hydrogel-tissue chemistry (which we first described in 2013, in a form called CLARITY; many variations on this theme have emerged since). In this approach, tricks from chemistry are used to build transparent hydrogels—soft water-based polymers—within cells and tissues. This physical transformation helps turn an intact structure like the brain (normally dense and opaque) into a state that allows light to pass freely, which in turn allows high-resolution visualization of component cells and their embedded biomolecules. All the interesting parts remain locked in place, still within 3-D tissue, evoking images of childhood treats—clear gelatin desserts with embedded bits of fruit that can be seen deep within.

A theme common to both optogenetics and hydrogel-tissue chemistry is that we can now observe the brain intact, and study the components that give rise to function, without disassembling the system itself, whether in health or in disease. Detailed analysis, always an essential part of the scientific process, can be carried out within systems that remain whole. The excitement resulting from these technologies (and diverse complementary methods) has spread beyond the scientific community—and helped give rise to national and global initiatives to understand brain circuitry.

By taking this approach—and integrating technological advances from other laboratories as well, in microscopy, genetics, and protein engineering—the scientific community has now obtained many thousands of insights into how cells give rise to brain function and behavior. For example, researchers identified specific axonal connections projecting across the brain (like warp threads embedded in a tapestry, entwined with countless other crossing fibers) through which cells in the frontal parts of the brain reach deep into regions that govern powerful emotions like fear and reward seeking, and help restrain behaviors that would otherwise turn these emotions and drives into impulsive action. These findings were made possible because specific connections defined by their origin and trajectory through the brain could now be precisely controlled—in real time, at the speed of thought and feeling, during the complex behaviors of animal life.

These deeply embedded axons help define brain states and guide expression of emotions. By grounding our understanding of inner states at the level of precisely defined physical structures in this way, we also obtain a concrete perspective on the past, on our evolution. This insight emerges since these physical structures were formed during our early development and infancy by the operation of our genes, and genes are what evolution has worked with in shaping human brains over the millennia. So our inner threads, in some sense, project across the time we have inhabited as well as across the space within us—a legacy anchored in humanity's prehistory, needed by our forebears to survive.

This connection to the past is not magical—this is nothing like "collective unconscious" communication, in the way that Carl Jung invoked mystical connections with distant ancestors across time—but arises from brain-cell structure, a physical heritage from our predecessors. Beings that by chance created the first early forms of these connections we possess (and study) today—with some variation from individual to individual—were likely to have survived and reproduced more effectively, and as a result, they passed the genes governing that brain-structure predisposition down to us, and to other mammals in the modern world. So we do feel what our ancestors likely also felt—not just incidentally, but at times and in ways that mattered greatly to them.

These inner states were bequeathed to us through the relentless will (and sometimes good fortune) of their survival—giving rise to humanity, with our feelings and our failings.

·

The promise of modern neuroscience even extends to the prospect of addressing human frailty and easing human suffering: from guiding therapeutic brain-stimulation methods with our newfound knowledge of causation (what actually makes things happen, with cellular precision) in brain circuits, to discovering the roles in brain circuitry of genes that are linked to psychiatric disorders, to simply stirring hope in long-suffering and long-stigmatized patients. So scientific progress has deeply informed clinical thinking—this is the value of basic research, nothing new though wonderful still—but my perspective is also inverted, in that clinical work has just as powerfully guided my scientific thinking. Psychiatry has helped drive neuroscience in return—and this is enthralling to consider: the experiences of suffering human beings, and thoughts about mouse and fish brains, are informing each other. Neuroscience and psychiatry are pulling together, bootstrapping, connected at a deep level.

In light of these developments over the past fifteen years, it is interesting to reflect on my initial self-perceived lack of personal connection with psychiatry. So profound was the impact of my first unexpected encounter on the psychiatry ward—the shouting, the fear, the vulnerability of experiencing a terrifying reality through another's eyes—that I sometimes wonder if I was by chance unwittingly prepared, already tuned, to be positively affected in a particular way by that moment, which for many people would have been, understandably, nothing more than a disturbing encounter. Personal inspiration (like scientific discovery) can come from unexpected directions, and so I now think of my course correction in that moment as a kind of parable about the perils of prejudgment, and the need for direct personal exposure to find true understanding of almost anything human.

There is another allegorical aspect as well, in which the story of optogenetics provides a lesson for the broader sociopolitical world on the value of pure science. Historical work on algae and bacteria dating back more than a century was essential for us to create optogenetics and gain insight into emotion and mental illness—but this path could not have been predicted at the outset. The story of optogenetics demonstrates, as transformations of other scientific fields have before and will again, that the practice of science should not become too translational, or even too biased toward disease-related questions. The more we try to direct research (for example, by concentrating public funding too focally in large projects targeted to specific possible treatments), the more likely we are to instead slow progress, and the undiscovered realms where ideas that will truly change the course of science, of human understanding, and of human health, will remain in shadow. Ideas and influences from unexpected directions are not only important but essential—for medicine, for science, and for all of us, in finding and following our trajectories through the world.

These days, I sometimes imagine seeking out that schizoaffective disorder patient, with whom I shared that heart-pounding first awakening, to sit down together for a quiet moment of communion—though it has been a long time. A receptivity to the improbable comes very close to the essence of illness on the schizophrenia spectrum, and so he might not be at all surprised to learn that his crossing of the threshold of the nurses' station that day could have helped in its own way to advance psychiatry and neuroscience. In a real sense, our conversation now could confirm for him, and for me, that despite the depth of his suffering, from some angle, some perspective, his warp aligns with all of ours, and blends completely into the shared tapestry of the human experience, within which he is no more ill than humanity itself.

CHAPTER 1

STOREHOUSE OF TEARS

The lines are straight and swift between the stars.
The night is not the cradle that they cry,
The criers, undulating the deep-oceaned phrase.
The lines are much too dark and much too sharp.

The mind herein attains simplicity.
There is no moon, on single, silvered leaf.
The body is no body to be seen
But is an eye that studies its black lid.

—From WALLACE STEVENS, "Stars at Tallapoosa"

The story Mateo told could only be held in my mind if abstracted, if I flattened the mental image like a collapsible gurney and slotted it in among all the others I had seen. It helped me to not acknowledge how long he had been suspended by his seatbelt in the flipped car, to not consider the feeling of helplessness as his family died around him, considering instead only an instant of time, a still.

Or instead, simplifying Mateo himself, I could reduce his dimensionality, the space he occupied—in my own mind compressing flat his human texture into a simple plane. Then I could bind his story together with others like it I had heard or seen—together they became like a stack of old newspapers, all borne together without individual features, fused in a rain of tears. In this way suffering could be summarized as a tractable single object of ten or ten thousand lives. *I don't know why I can't cry*, he began, and when it was all told and bundled, it was not more, or less, than any other ending of a human world.

There is no formal protocol in medical training to protect a doctor's exposed heart in these particularly devastating moments. Physicians and nurses, warfighters, crisis workers—they all come to learn defenses on their own, in order to live among the extremes of human suffering. It is not just the magnitude of the pain but also its incessance—the unrelenting descent into the abyss, day after day, year after year—that without some safeguard would be unsustainable.

Our natural impulse is to connect deeply and broadly with someone in a state of personal loss, to try sensing inside our own minds a full and complex representation of the other, to fully understand what the tragedy means. But in the extreme context of horrific suffering, it can be helpful to instead narrow our perspective to preserve empathy, finding a point to experience within the broader tapestry of the patient's life, focusing on one spot of interlocking threads creating local shape and color.

It is important to know that the full perspective is available, but feeling fully does not make sense of tragedy—and depths of emotion do not seem to help us with precision tasks in the agonal moment, whether performing an artful lumbar puncture of the spinal column or a difficult psychiatric interview to elicit inarticulable feelings. Our perspective widens when it can, sometimes without warning—on the drive home, or among our children with a sudden sob. Until then, out of view but always accessible are the trajectories of the patient's threads with all the scope of life and its dreams, from their anchors and origins, through the journeys and relationships that came together in that instant of catastrophe and strife.

Each tragedy is still felt intensely, and each suffering human being is held in the heart carefully no matter how many more come over the years—every stunned and bereft father after a car accident, every mother struggling to form words as she hears her child's brain cancer diagnosis. And care is needed; when the cases accumulated are still few, early in the life or training of a doctor (and sometimes still later), a single experience can storm and overwhelm the inner self, in the part of us that sees and feels representations of human beings, textured images of valued others, positioned carefully like tapestries in the innermost fire-lit halls, the hidden spaces of the self. In the keep, if we were castles.

•

I should have been better prepared, but there had been no warning that the keep was vulnerable. Until meeting Mateo (in my role as the on-call senior psychiatry resident who was summoned to the emergency room to evaluate him), I had not been hurt badly by my own empathy for years, not since I had been a raw and young medical student. But all had been different back then—my feelings were mostly only feelings in medical school, not feelings about feelings, a safer form they achieved later. And as a medical student I had been more vulnerable: coming of age in the halls of medicine, not yet able to give orders nor to prescribe, and still learning the language of the field though raising a child of my own as a single father in the world beyond.

On that night that hurt me first and deepest, years before meeting Mateo, I was the medical student on pediatrics at our children's hospital, on a call night that had not been too busy. My first task for the evening—a brief prelude for what was to come later that night—had been admitting and taking the history from a cystic fibrosis family. The patients were three-year-old twins brought in together for respiratory distress. It was hard for the children to breathe.

The family was well known to the service, as we say. There had been many past admissions to the hospital, and the parents were pros at the process—so much so that they answered my questions as soon as I began to ask them, so much so that they were in the process of divorce.

They had found, with the birth of the twins, what seemed to be a hidden flaw in their union. In most cystic fibrosis families, parents have no symptoms themselves, but each carries one copy of a mutant gene. Mammals have two copies of almost every gene, so often if a single gene is damaged, no ill effects are seen; the other copy can allow a healthy life.

Mother and father are healthy carriers in cystic fibrosis, with the load carried not usually known until their child is born with a far heavier burden—both damaged copies of the gene. The math is simple, and I learned that night that the parents, in their own relative youth, had arrived together at a seemingly simple and practical decision to split and remarry, each to seek a noncarrier and the promise of greater health in the family. But in the meantime, before this defiance of the blind forces of population genetics could play out, I had to work through the mucousy tumult of the sick and screaming twins, patiently building my inventory of facts, collecting their medical history above the din, and completing the admission.

By midnight, quiet had finally been restored when we got wind of a late emergency transfer arriving from an outside hospital: a four-year old girl, Andi, with a brainstem finding.

I would carry this one, and what came next, with me for years: a deep gouge, perhaps deeper than I know still. Perhaps it went all the way through. I helped admit Andi to our inpatient service—she was charming and moony in a high ponytail, kneeling on her hospital bed, arranging her dolls around her, eyes just a little crossed up, one turned a bit inward. It had been almost beneath notice as she played catch with her family earlier in the evening, a detail almost lost in the special thrill of staying outside later than usual—just a little double vision in the twilight, and then just a twinge of worry.

I found myself in deep very quickly, even though I was the least important part of a small collection of people brought together by the case, all crammed together in the team workroom of the inpatient unit. I had started the meeting leaning against a wall, and it immediately became impossible to consider sitting down, or even to shift weight from one cramped leg to the other, as the emotional impact of the scene manifested before me. I was frozen in place until we were most of the way to dawn.

A lone gray rectangle of film, the brainstem scan, had been brought by the parents—clutched and hated, their ticket to ride from the outside hospital deep in the valley. They had borne the scan through the evening to the windowless team room, and it was now jammed up on the lightbox, a backlit gray requiem. Andi's parents, eyes red with latent tears, were across from me—seemingly transposed into a separate space, somehow alone in the crowded room. The doctor of record—the pediatric neuro-oncology attending—was positioned immediately to my left, sitting down and leaning forward. He had been paged and brought in, as late as it was, not to perform a procedure, nor to make a clinical decision—there was nothing to do that night—but to explain the conclusions from our physical exam, and from the reading of the film, to the family.

Words were the only tool for the neurologist that night. He leaned forward for hours, not relaxing back once, without a glance at me or anyone on the team, his words only for two in the crowded room, for mother and father, for two alone, all through the night.

The double vision was not a mystery to us. There was a finding on the scan. A shadow had fallen across her pons.

At the stem of the brain, at the base of the skull, there is a bulge of cells and fibers called the pons—dense and vital, connecting all that makes us human in the brain above to the spinal cord and nerves that leave the skull below. If within the pons—Latin for bridge—a disruption has appeared in the path of fibers passing through, physicians can see this happening without a CT scan or an MRI: no medical imaging, just human imaging, just looking into the eyes of the human being.

Andi's eyes were crossed, but with only one turned inward, toward her midline—because a tiny muscle on the side of her left eye globe (called the lateral rectus, for its task of turning gaze out laterally, tracking a baseball thrown wide) had failed. The fine shred of muscle was no longer receiving instructions from the brain; its dedicated channel of communication, its nerve, had fallen silent.

The sixth cranial nerve (of the twelve that emerge from the skull) is called the abducens, and is beloved by confused medical students for its unusually straightforward trajectory (compared with the other winding and branching and crossing cranial nerves). Abducens, the simple sixth, is one nerve serving one muscle, the lateral rectus muscle with its one job—abducting the eye. The abducens lies entirely on one side of the brainstem, its threads coursing deep through the heart of the pons, carrying out their singular duty.

But now, tonight, abducens was playing another role, reporting back on something unnatural in the brainstem, telling a tale of a thing gone wrong—and on the film, confirming the diagnosis, we could see that a form, a darkness, lay across the bridge. The neuronal threads on one side of the bridge were disrupted, and so the eyes no longer turned together, were no longer aligned toward their common goal.

This coordination between eyes is a lovely thing when it works, when they join to face the world together in primates like us. Both receive the same instruction from the brain, to follow the ball thrown from Dad in the darkening chill. But the two eyes, each with slightly different shapes and angles, are not connected to each other. Much has to be perfectly tuned for the two to move together, to not create double vision with offset views of the same scene.

This challenge is especially pleasing to bioengineers, as a paradigm of the need for design. Such synchrony and symmetry in biology, when achieved, imply trust, and truth, and health. Two sensors, two eyes, are balanced together on the finest edge of time. There are always failures of communication in biological systems—noise, variance, chaos, there is even sometimes profit in deception—so every system needs feedback to check and to calibrate. Early in life, before we become aware, double vision serves as that error signal sent back—and then our brains mend the error, adjusting the instructions transmitted down cranial nerves to the eye muscles, aligning and tuning with care, until the offset disappears, and we see the world as one.

The world becomes whole, until a falsity returns in some—it was here tonight in this little girl, and here it could never be trued. When a member of this pair slips infinitesimally, an intruder is revealed, and a sickness known; the nerve fibers through the pons become more and more disrupted, as the shadow spreads. Here it could have been no other cranial nerve, it was abducens, sixth nerve of twelve; in this brainstem cancer, it is always the sixth: alert and direct, a border division reporting without fail the first faint hoofbeats of invasion.

The attending was careful to give no firm prognosis that night, though I had paid enough attention in class and on rounds to know that a death march had begun. This was DIPG, a diffuse intrinsic pontine glioma, and she would have six to nine months of life. Her parents were sensing this but not knowing, not getting numbers but feeling the unraveling as a new reality took hold, as a fibrous invader insinuated itself throughout their inner worlds, entangling their every thought and sensation down to the feeling of breath, of life itself. Their words were dry, and strangled, and dragged thick from their throats.

I knew what was far worse, what they could not suspect then. I knew what kind of death would come. Within a few months Andi would be unable to speak, unable to move—paralyzed with her eyes wide open, still as bright and alert and perceptive as she was that night. Locked-in—a state of straight nightmare—as the bridge fell, as all her pons gave way.

So quickly, all had turned. Just a visit to a local doctor, for double vision in their little girl, on a weekday night. My own first child was nearly the same age, almost four, though it was hard to consider that fact in my mind for more than a moment. That night, every time that thought came forward it was shut down fearfully by some other process inside, with the feel of a heavy gate slamming shut. A crude and immature defense, as I recognized even then—don't look, don't connect—but temporarily effective.

In the days that followed, a new kind of grief made itself known to me. As I learned to hold open that gate just a crack, just enough to let a little light pass through, just enough to see Andi's connection to my son—and to glimpse in this way, so far beyond what I could fully imagine, the grief of her parents—as I did, angry tears would come as I felt a nonsensical anger at the disease entity, a rage that DIPG was part of our world. There had to be hope to defeat this malevolence; there had to be hope for Andi.

At my lowest, an unexpected thought appeared in my mind, seeded by this child, nurtured by this anger—that some could live this way, but I could not. I could not last in medicine, not for a lifetime of this. I would retreat instead, to havens dumb—to the lab, I told myself—the harbor of science I knew so well, a place where no girls die.

Yet with time this storm, of grief and anger and false hope and retreat, spent its energies. New experiences surged to the fore of thought and feeling. I healed—but still immaturely, walling off the hurt gradually, as an abscess forms to seal off an infection. With time I just stopped hoping; all I could think was that the world needed more than hope from me.

There was nothing to do for Andi. With DIPG there was no surgery that could find its way safely in and around the fibers of life and breath and motion in the pons, and no chemical or radiation had lasting power. I was no more able than her parents to protect her from what had come, cloaked in the phantasmal darkness of the brainstem, in stealth under skull and skin, and under the thin membrane still cradling her brain. Pia mater, we call that membrane. The loving mother.

When I stopped hoping, my tears stopped. I focused outward, on the homely details that form our lives. I rotated off pediatrics, never seeing Andi again. It was not bearable but borne, her end is known but not seen, and she is kept in me now.

•

Still today, those feelings reach throughout almost every part of me—stopping now before tears. That inner state is always there, ready to return, although the emotion is gentler and more complex now; the world has changed, and I have changed. There are more representations of others deep inside me, interconnected with Andi, and supporting her.

Those memories are now also textured with the progress of science, and with development of the method of optogenetics that allowed me to peer into the brain's inner workings, to explore how internal states of emotion are constructed at the level of cells, and to test how these elements of construction matter. This method works by re-creating a part of one organism's design within another living being, in a way that allows this new part to persist in the recipient, and to be integrated into the whole. The part, a gene, then influences action in the host by providing a new code of conduct—just as an insight or novel experience can do.

It is common in biology for one organism to cross the boundary into another—sometimes spontaneously, sometimes by design. It can be a single cell that comes across the frontier, bearing only life's universal essence—DNA, a genetic program, a living acid—within a thin lipid cover, borne lightly across the border on this frail life boat. This is the story of life on earth, and it happens every which way. Especially when the distance is far and barriers formidable, the opportunity is great for those on both sides of the border.

Every plant and animal on earth, and thus every human being, owes its life to such travelers from a foreign kingdom—members of an ancient class of microbes called archaebacteria—who brought the mysterious skill of using oxygen for energy when they traveled into, and dwelt within, our cellular forebears more than two billion years ago. Were the travelers invaders, entering by breaching the barrier, seeking to consume and destroy? Or were our own genetic ancestors the aggressors, hunting and absorbing, enveloping the smaller free-living supercharged oxygen burners?

In the end it is topology that matters, not intent. What is of consequence is that an entity has passed across the border. The migration is risky for both, but when the larger organism learns from the smaller, retaining rather than destroying, then the perilous border crossing can instead give rise to a new kind of being. In the case of our lineage, we were brought the breath of life itself.

Suddenly living together, the two kinds of life had to coevolve if they could, mutually accommodating each other's limitations and oddities. There was time enough to sort it out, hundreds of millions of years—as long as the union was not immediately catastrophic—time for the new joint being to evolve, following the same Darwinian selection rules that gave rise to life itself, and that had allowed each part, each partner, to arise alone at first.

Subcultures can be preserved in union. The small oxygen burners became our mitochondria, the energy factories for each cell. So ancient in origin that they use a different dialect of life's DNA code, they have kept their mother tongue for private use over billions of years of living together with us. At the same time the microbes adapted themselves to our culture in countless other ways for the shared goal of survival. And we adapted as well, coming to need the burners as absolutely as they need us—they have become part of us now, and we will never be apart again.

These microscopic migrations, from microbe to animal, or microbe to plant, are globally significant. These transitions can change the overall flow of energy on the planet, sun to plant to animal, and so change the landscape of earth. Many times these migrations have happened, and some have persisted. Though the success rate is infinitesimal, the universe has had billions of years to work with—and over that time, low probabilities become certainties.

But in the last fifteen years, taking a shortcut through human hands with optogenetics, microbial DNA has yet again returned to animal cells. The microbial genes are targeted not to our bodies but to cells of animals in the laboratory—and not spread across kingdoms through chance encounters, but rather guided by scientists who accelerate this information transfer, spanning immense genetic and conceptual spaces, bridging the branches of the tree of life.

Today, seeking to control brain cells with great precision—in order to discover how the brain's wonderful workings arise from pulses of electrical activity in cells—we supplant the random hand of evolution. Not wanting to wait a billion years, we put certain genes from another ancient DNA data stream—still persisting in microbes of the natural world—directly into mammalian neurons. We do this to take advantage of a distinct alchemy that this different class of microbes developed—turning not oxygen, but rather light, into energy and information—via specialized genes (called microbial opsins) that allow conversion of light into ionic current flowing through the surface membrane of a cell. And ion flow, the movement of charged particles, happens to be the natural signal for activation and inactivation of neurons.

Most neurons don't normally respond to light like this, but all they need to do so, it turns out, is a single foreign gene—a microbial opsin. And with a few more components from human experimentalists—genetic tools to put the opsins into specific kinds of cells to be tested (so only those respond to light while all the others remain unchanged), and special light-guidance tricks for sending in laser light (via fiberoptics or holograms, to bring the light only to certain cellular structures)—optogenetics was created.

In this way we can elicit electrical activity directly in neurons by sending light from a distance into animals carrying out complex tasks of life, just as a conductor brings forth music from the orchestra. If brain function is the music—sensation, cognition, action—then brain cells are musicians ten-millionths of a meter wide, numbering in mammals from millions to billions. Optogenetics is the conducting of activity in neural circuits using light, eliciting a music of the natural world, an animal performing to its design, with form and function together arising from the individual cells and cell types in the brain.

Optogenetics brought together these two patients of mine—a little girl and a young man, Andi and Mateo—linking like two notes of a minor chord these two human beings who had come to me for help, each with a disease that had disrupted a different natural inner harmony at nearly the same tiny spot, deep within the most ancient region of the mammalian brain.

•

"Why am I here tonight?" Mateo asked. He took his glasses off and set them carefully on the gurney. "Because I don't know why I can't cry."

Looking at his hands, open in his lap, he considered each palm in turn, seemingly puzzled by its emptiness. Then his eyes came back up to mine, and his story began to slowly drain out, passively, by force of gravity.

He had been brought to the ER by his three brothers, who were surging about in the tiny waiting room down the hall. With my first take, on stepping into the room, he had seemed childlike—just twenty-six but somehow looking even younger, with smooth skin and rich brown eyes framed by thick black glasses, sitting all alone in Room Eight. He looked as though he had lost his backpack, or perhaps was worried about his homework. And yet that impression lasted only an eyeblink.

Eight weeks earlier, he told me, his wife of a year—his pregnant bride—had been crushed and killed in their car. She was stolen from beside him late one night, as they drove in darkness on a country highway. They were returning from a weekend getaway at a bed-and-breakfast in Mendocino, when a white panel van had cut across their lane.

Mateo couldn't brake in time, the van loomed, death dilated. In the last moment he fought as hard as any mammal could have. He jerked the wheel hard left, and their little car flipped into the median, finding a small tree stooped there that had been quietly waiting fifty years for this moment. They hung upside down for an hour, Mateo untouched, trapped beside his wife's broken body, the young family swinging quietly in their seatbelts—along with the little one too, deep within her, cooling slowly along with her, unsafe in her soft embrace.

He stared at the wall now, arms empty. Two months later, there was still visceral horror in his heart—but also a relentless dry isolation. *I don't know why I can't cry.* Following his lead over the next hour, I asked for more, learning about his life, his calling, his immigration from Barcelona. He was an architect and a lover of chess; he cried on his wedding day seeing his bride come around the path in the outdoor garden, and again soon after—when he learned she was pregnant.

Here was a man whose inner self, his emotions, had been projecting out into the world—but his dimensionality was now reduced. Even his phrases were flat and colorless. He seemed set aside, set apart in time, sighted in one direction only. When I asked about his plans, there was only nothingness. He could not see even a few minutes into the future, which was invisible, impossible, a featureless white wall.

As blank as his future was, Mateo still suffered in great detail the complexity of his past. A particular thread haunted him. Spinning, churning, his brain was focused on a moment from years earlier, when he had struck and killed a highly intelligent mammal, a raccoon—also on the highway. He had been alone, barreling down the broad I-280 in the early morning half-light, and the raccoon was there in the fast lane, frozen, looking back at him. He did not turn then, confident in his mass, his big machine, aware of the risk of swerving at that speed—just rode it on through, for the good of all, and especially since the call was his to make, the life his to take. The impact had been only a moment, a thump. The family was back home in the den waiting for warmth and food but it would not come, not now, not ever. His car rode on, rode through, and brought Mateo—Mateo alone—home.

He earnestly tried to understand what had happened. How much of what he had done earlier mattered later? He played and replayed the moves of his life—did he have to swerve too hard with his wife because he had not earlier, because he had not sought to spare the life of another? His mind was occupied with unpiling past decisions from each other, dissecting moves and links and connections—but it was all now just insoluble rumination. He had left himself alone on the board, a lone king, pointless, in stalemate. He wanted to pound the earth, to demand of God to know why he was still here. *I don't know why I can't cry.*

•

Unexpected absence of tears in a bereft newlywed, unexpected presence of tears in a young medical student—and all the other times when tears surprise us: this complexity and subjectivity might seem inaccessible to science. To even approach understanding of these mysteries, a scientist might first seek a way to reduce, to simplify—to find a viewpoint that cuts out the subjective and leaves something measurable. Yet here, the whole and the essence seems to be subjectivity.

Such a conundrum need not be the end of seeking to explain; most modern fields of inquiry have been, at some early point in their history, not readily welcomed into the conversations and canons of science. New ideas are often consigned to dwell on the outskirts for some time, yet eventually develop into acceptable scientific discourse, as long as something interesting can be measured consistently. For example, in one of the most recent and spectacular of these transitions in science, we now know for certain whether our species, *Homo sapiens*, mated with prehistoric hominins, Neanderthals, with whom *H. sapiens* co-existed in Eurasia for many thousands of years. The subject of speculation and romantic fiction only a few decades ago, this question gave way in recent years to unequivocal factual knowledge. Not only do we know that Neanderthal interbreeding happened, but we also know exactly how much of modern Eurasian human genomes arose from this interaction—about 2 percent. This transition from fiction to science was due to the birth of a new kind of measurement—in fact, a new field, named paleogenetics—that arose from the marriage of technology (for DNA sequencing from fossil bones) and human curiosity (embodied in the work of a few pioneering modern genetics labs).

Questions about who we are, and the nature of our origins, are better posed now having taken the measure of that 2 percent. But still to be explored are many details (with some accessible by DNA sequencing) of the drama and tragedy swirling around the African and Eurasian cauldron of inter-hominin mating and extinction up to forty thousand years ago—only fourteen hundred generations past, when the last breath of a last Neanderthal was weakly drawn and without a sound exhaled into the dank air of a hidden cave, alone in a final redoubt near the coast of Iberia.

And having taken this measure, paradoxically the mystery of humanity's long march is not diminished by pairing a question with an answer, and finding a number like 2 percent. Scientific knowledge expands the scope of human imagination so that fantasy can launch from deeper pilings of understanding in the bedrock of the natural world, and reach further. Now on this same trajectory are other recent arrivals to hard science—even the inner states of the mind, like anger, and hope, and psychic pain—states we had known before only by our own experience, as they come uninvited like light and weather, as storms and dawns and creeping dusks.

•

The scientific process almost always begins with measurement—and inner states, though subjectively experienced, can have measurable manifestations. As optogenetics experiments have shown, these manifestations may find physical form arising from the trajectories of axons, the threads that form the three-dimensional tapestry of the mammalian brain. Exploring the threads of anxiety was an early example of this kind of scientific advance.

Anxiety is a complex state, with features we know from introspection: changes in bodily function (heart rate driven high, breaths coming fast and short), changes in behavior (apprehension and skittishness— avoidance of risky situations even if there is no immediate threat), and finally, subjectively, a negative or aversive internal state (feeling bad, one could say).

Such utterly distinct features would, perhaps, have to be generated by correspondingly diverse cells in the brain. Optogenetics (along with other methods) illuminated how this complex state—so familiar to most of us—could be assembled and disassembled by different cells and their connections across the brain. For each of these components of anxiety (breathing rate, risk avoidance, and that unpleasant internal feeling), different axonal threads that could be responsible were discovered, and accessed, and controlled independently with optogenetics. Here is how it was done.

Imagine a spot deep in the brain, a single anchor point, with many threads radiating out as if from one beam of a loom to another, each stretched to connect with a different target location across the brain. This is not unlike how outgoing neural connections (in the form of axons) venture forth from a single anxiety-control region, a deep brain structure called the amygdala—even more precisely, from an extension of the amygdala called the bed nucleus of the stria terminalis, or BNST.

These threads stretch, and dive, and go deep to find the cells needed to make all the parts of anxiety. One even goes to the pons, to the spot of Andi's shadow.

Amid all the interwoven intricacy of the brain, how can we know this—that these threads in particular actually matter? Here is where we can introduce the genes from microbes, to provide a new logic to each thread. Into quiet darkness under the skull, we deliver a new code of conduct from a foreign being. We teach one connection, and then another, and then another, to respond to light.

We borrow a lone microbial gene from single-celled green algae; this gene is just DNA instructions for making a light-activated protein, called a channelrhodopsin, that lets positively-charged ions into cells (which is an activating stimulus for neurons, making them fire away, and broadcast their signals). We deliver this gene into the mouse BNST, smuggling it in via a virus that we have chosen for its aptitude at bringing DNA into mammalian neurons. The cells in the BNST, having thus unwittingly received the algal gene, begin to produce the algal channelrhodopsin protein as instructed—duly following the DNA blueprint, the assembly manual written in the universal genetic script of life on earth.

At this point, if illuminated with bright blue light, each of these BNST cells would fire action potentials, the spiky signals of neuronal electrical activity (and the light would be easy enough to provide, with a nearly hair-thin fiberoptic placed just right so that laser light sent through the fiber would shine forth into the BNST). This would be a totally new capability, a new language taught to animals by the algae, with our help. But in these anxiety experiments, we actually don't bring in the light just yet. We wait, and an even richer language emerges.

Over several weeks, the channelrhodopsin protein (which we have linked to a yellow fluorescent protein, so we can see where it is produced and track its location) fills up not only the cells in the BNST but also their threads as well, the axons, which after all form part of each cell. Every neuron in the BNST is built with its own outgoing axonal connection, and different cells send their threads to different parts of the brain. After several weeks, radiating out from the BNST like rays of the sun, yellow streaks of the channelrhodopsin-linked fluorescent protein extend across the dark secret interior to all of the brain regions that the BNST speaks to, that need to hear a message from this anxiety center.

Now the new capability becomes clear. A fiberoptic can be placed not in the BNST but in an outlying region—indeed, in each of the different regional targets of the BNST across the brain. Laser light sent through such a fiberoptic can then do something quite special. The only light-sensitive part of each target region, upon which a yellow thread lands—for example, the pons—is the set of axons from the BNST to that region. And thus light delivered (to the pons, in this case, that deep and dark pedestal of the brain stem) activates directly only one kind of cell in the brain—the kind that lives in the BNST and sends axonal connections to the pons. A single kind of thread in the tapestry, defined by anchor and target, picked out from all the interwoven fibers, is now directly controllable by light.

When this is done in mice, a connection from the BNST to the pons—home of Andi's abducens and also home to a subregion called the parabrachial nucleus, which is involved in breathing—was discovered to control respiratory rate changes when activated, but had no other effects that we could see. Stimulating this pathway optogenetically affected breathing rate as seen with changes in anxiety, but interestingly had no effect at all on the other features of anxiety—the mice showed no change in risk avoidance, for example.

Instead, risk avoidance was controlled by a different thread—the connection from the BNST to another structure called the lateral hypothalamus (not nearly so deep as the pons). Activating the cells of this pathway, using optogenetics, changed how much a mouse would avoid exposed areas of an environment (the middle of an open area—the most risky place to be, if one is a mouse and vulnerable to predators), without changing anything else (for example, no change in respiratory rate could be seen). Thus a second feature of anxiety was picked out cleanly, defined by another cell type, and we begin to see that different parts of inner states are mapped onto different physical connections.

What about that third feature of the anxiety state, feeling bad? We call this *negative valence*—and its opposite is *positive valence* (a good feeling, like that of sudden release from anxiety, which feels to be much more than just absence of negativity). At first glance, this aspect seems hard to assess, especially in a mouse that cannot use words—and perhaps it would be difficult also in people, where even words are imprecise and not fully trustworthy. But still such an internal state—however subjective, however experienced by a mouse—can have external measurables.

In an experimental test called *place preference*, an animal is free to explore two similar connected chambers—just as if a human being had free rein to explore a suite of two identical rooms in a new house. If the person in such a situation were caused to feel acute and intense positivity inside (like that inner rush of a wild kiss returned, somehow feeling this without the kiss itself) immediately with every chance entry into one of the rooms, which terminated immediately upon each exit from that room, imagine how quickly that person would simply choose to spend every possible moment in one room. A single measurable—choice of the room with that feeling—reports to an observer on the hidden internal state. The observer cannot conclude precisely how it feels, of course, only that it is of positive valence—and a panel of additional tests can confirm that interpretation. Negative valence is tractable too. If the caused feeling is in that direction instead (inner negativity, perhaps in the same sense as the sudden crushing loss of a family member), then avoidance, instead of preference, becomes the measurable.

Valence can in this way be explored in animals, where optogenetics provides a means to instantaneously test the impact of activity in specific cells and connections across the brain. In the mouse version of place preference, the animal is given free rein to explore two similar chambers of an arena—first without optogenetic input. Then, a laser is brought to bear, set up so that light is delivered through a thin fiber-optic to the brain automatically, but only when the mouse happens to be in one of the two equivalent chambers (say the left-hand one). If there is an aversive, or negative, quality to activity of the particular optogenetic target of the moment (the specific neuronal thread that had been made light-sensitive in that animal), the mouse rapidly begins to avoid the left-hand chamber. The mouse, it seems, does not want to spend time in places associated with negative experiences—nor would we. Conversely, if there is a positive internal association, the mouse will spend more of its time in the chamber linked to light—revealing place preference.

Which thread deep in the brain, winding its way out from the BNST, governs this important feature related to anxiety—that of positive or negative valence, perhaps corresponding to the subjective feeling of our own internal state? Surprisingly, neither of the two other connections mentioned so far, to pons or to lateral hypothalamus, governs this behavior from the BNST. Instead, this job is handled by yet a third projection, from the BNST to another spot that lies deep, almost to the pons but not quite—the ventral tegmental area, or VTA, where neurons live that release a small chemical neurotransmitter called dopamine. This group of cells encompasses its own diversity of roles and actions, but overall is intimately linked to reward and motivation.

Activity along the other two projections, to pons and lateral hypothalamus, does not seem to matter at all to the mice—with stimulation affecting breathing and risk avoidance but not positive or negative associations, at least as far as the place-preference test is able to report. Even more strikingly, the third thread, to VTA, does its job of place preference in mice (and could therefore implement subjectivity in people) without affecting, in turn, the other features—breathing rate and risk avoidance. Thus a complex inner state can be deconstructed into independent features that are accessed by separate physical connections (bundles of threads defined by origin and target) projecting across the brain.

Not limited to the study of anxiety, this same approach turned out later to be applicable to mammalian behaviors in general. Even the complex process of parenting, in the form of intimate care of mammals for their young, was soon deconstructed into component parts, mapped onto projections across the brain. This discovery came from another group of researchers five years later, using the same optogenetic toolkit and projection-targeting approach. Many mysteries remained for anxiety, of course. For example, this deconstruction of the anxious inner state does not answer (though it strongly frames) a timeless puzzle— the questionable value of having positivity or negativity of internal states at all. The clean separability of place preference from risk avoidance highlights a deceptively simple question: why must a state feel bad (or good)? If behavior is already tuned and controlled appropriately for survival—if risk is already avoided, as dictated by the projection to the lateral hypothalamus—what is the point of the preference, or subjective feeling, provided by the connection to the VTA?

We think that evolution by natural selection works with actions taken in the world—that what is actually done, rather than felt, by an animal affects its survival or reproduction—and so perhaps how the animal feels inside, or how we feel inside, should not matter if action is already addressed. If the mouse is already avoiding the risky open space, as it should for survival and as governed by the BNST-to-lateral-hypothalamus thread, with no positive or negative association at all, then what is the purpose of the separate VTA thread and its associations? Feeling bad seems gratuitous—and more than that, a vast and unnecessary source of suffering. Much of the clinical disability in psychiatry, after all, arises from the subjective negativity of states like anxiety and depression.

One reason may be that life requires making choices between utterly distinct categories that cannot be compared directly. Subjectivity—feeling good or bad, for example—may be a sort of universal monetary instrument for the inner economy of the brain, allowing positivity or negativity from diverse pursuits, from food to sleep to sex to life itself, to be converted into a single common currency. This arrangement would allow difficult category-crossing decisions to be made and actions chosen—quickly and in a way best suited to the survival needs of the particular animal and its species. Otherwise, in a complex and fast-paced world, wrong calls will be made: freeze when a turn is called for; turn when a stop is needed.

Perhaps these conversion factors are something that the evolution of behavior works upon. Relative value (in the common currency of subjectivity) assigned by the brain to different states will inevitably determine consequential—indeed existential—decisions made by the organism or the human being. But these currency conversions also must be flexible, needing to shift over life and over evolution, as values change—and this flexibility could take physical form, such as changes in the strength of threads connecting to valence-related regions like the VTA.

The insight that the optogenetic study of anxiety brought me was that subjective value (positive or negative) and external measurables (breathing—or perhaps crying) could be added to or subtracted from brain states with eerie precision. But this understanding came years later for me, well after Mateo had entered and left my life. At that time in the emergency room, I had no way of knowing that separability of one element of an inner state could be so precise, nor that this could happen as a result of the physical form taken by that element (involving electrical activity traveling along a connection from one part of the brain to another). Seeing Mateo, I had no framework to understand how he could be unable to cry, as he normally would—while lacking none of the other human elements of profound grief.

．

Still today, deep mysteries of our inner states remain outside the reach of science. It can seem in poor taste to study love, or consciousness, or crying. With good reason—if no objective and quantitative tool (like paleogenetics for insight into Neanderthal prehistory, or optogenetics for discovering principles of brain function) yet exists, answers may lie beyond our grasp.

In the case of crying, a biologist should presume that if fluid is ejected from a duct like those connected to our tear glands—with precise timing, and in consistent contexts, for individuals of the species—there is likely an evolutionary reason, and the matter is suitably objective for science. If changes in duct performance come along with a strong feeling, a subjective internal state, then the combination of subjective and objective should intrigue a scientist, a psychiatrist, or any student of the human mind and body.

Crying is significant in psychiatry. Our patients are experiencing extreme emotions, and we work with these emotions—with their articulation, recognition, and expression. We are experienced in seeing the less genuine tears as well, across a spectrum of deception from the mildly suffering and modestly manufactured, to fully professional manipulative tears. Yet little is known in the science of emotional tears, such as it is.

Emotional crying cannot be studied well in animals. Pure emotional tears, as we experience the phenomenon, are not clearly present elsewhere even among our close relatives in the great ape family; the reason, if any, is a mystery. Tears are powerful for driving emotional connection; it is known that digitally altering tears in human face images will cause significant changes in the sympathy, and the impulse to help, elicited in viewers (much more than for altering other facial features). But we are no more social than our cousins—the chimpanzees, the bonobos—and still, using the mystery of tears, we alone cry, and cry alone.

We display our inner state with this odd external signal, with or without audience, not requiring volition or intent, just broadcasting feeling to all observers and ourselves. But it is not only our great ape kin who seem excluded—even many of our own *Homo sapiens* do not shed tears emotionally, and so dwell just a bit apart. This separation may be one-way—those whose bodies do not generate the language still can understand and respond to emotional tears in others—but missing even this one part of the conversation may come at a cost; non-crying people have been found to show reduced personal attachment patterns, though it's not known if this association is more due to life experience or innate predilection.

The fact that this involuntary signal of emotional tears is absent from some human beings and from our close nonhuman relatives may be a sign of an incompletely established evolutionary innovation— perhaps because its value is not universal even today, or because it was a recent experiment: an accident still in the process of manifesting fully in the human family, or failing to do so. Every innovation in evolution is accidental at first; perhaps emotional crying came about initially as a chance rewiring of axons. Like the various projections from the BNST, all axons are guided during brain development to grow in specific directions by a vast diversity of path-setting molecules as strong as thread guides on a loom—tiny signposts that send a slowly growing bundle of axons on to the next brain region, or turn it back if it has come too far, or send it across the midline to the other side of the body. All of this, like everything in biology, was built by chance mutation over millions of years and so can find its way to new functionality by chance mutation as well.

A mutation at any of these steps—in any gene guiding the positioning of path-setting molecules, and so redirecting those long-range threads, the axons, across the brain—would be all it would take. Fibers coming from the emotional areas of the brain would change course slightly, and then there would be born into the world a new kind of human being, with a new way of expressing feeling.

Such innovations would have the potential to open up a separate channel of communication—with remarkable efficiency, considering that the actual biological changes needed to implement this innovation would be very small: one set of axons missing one guidepost and traveling just a bit too far during development. As almost always is the case in evolution, the key players would have been already present, just needing to be taught a new rule, and so create a new role. In this case, the relevant axons—such as those already traveling from forebrain areas like the BNST to deeper and ancient brainstem areas like the parabrachial nucleus for breathing changes—would have been just partially rerouted to a new destination.

Near the parabrachial nucleus lie the origins of two cranial nerves—not only the sixth, named abducens, the one Andi's cancer disrupted, but also its neighbor the seventh, called the facial nerve. All of these structures, just collections of cells, sixth and seventh and parabrachial, jostle together in a small spot in the pons, huddled on the bridge from brain to spinal cord. But the new target here, for tears, would be the seventh-nerve cells. For emotional expression, the seventh is a maestro in itself, much more intricate and multipurpose than abducens, sending and receiving rich information streams to and from many muscles in the face and many sensors in the skin. The seventh, the facial nerve, is grandmaster of facial expressions but also of the lacrimal gland, the storehouse of tears.

The lacrimal system likely evolved for flushing irritants from the eye, washing away particulate nuisances. Now, with almost trivial rewiring, it could be involuntarily recruited by a flood of emotion, perhaps alongside other fibers reaching to the breathing centers—the parabrachial and beyond—wrenching from within us the cathartic, diaphragmatic contraction of the sob. When the first human being who had this mutation cried, and perhaps even sobbed, what might have been the effect on those nearby—friends or family or competitors—who had never seen this before? Communication through the eyes would have long been important, always a focus of attention for human beings—eyes are rich in information and constantly accessed—so the innovation would have landed fortuitously in real estate of high value for sending signals. But there might have been at that moment no understanding, and no emotional response, to the tears—just attention and interest to the unusual and salient signal. Full meaning and value, for survival or reproduction, may have taken generations to evolve.

If there is evolutionary significance of crying at all, then the times when emotional crying occurs may provide clues. Largely not voluntary in humans—this is a signal far less under our conscious control than, for example, smiling or grimacing—crying is a mostly honest journalist, reporting for some reason on a kind of feeling. Scholars have focused on its value for social communication, but emotional crying also happens, and feels important—even productive, addressing some need—when we are alone.

Given all the risks of revealing true feelings (and all the individual benefits of successfully conveying false feelings, for beings in complex social environments), the poor controllability of this emotional readout seems at first a handicap rather than an advantage—something to be selected against, rather than selected for, at the level of the individual. Signaling to self, or to others—either way, it is interesting that this signal has remained mostly involuntary, and thus mostly true.

Is crying still evolving, under selection pressure, to either escape or come under our volition? We might eventually control crying as easily as smiling, unless its involuntary nature is more useful than individual advantages that would come from voluntary control. And right now this truth-signaling property is known broadly to the species in some sense, programmed in human observers to be of greater impact than more easily gamed facial expressions like smiling—thus heightening its effect on others, perhaps drawing fellow human beings closer for bonding and support, perhaps in times of true and desperate need.

In that case, a sort of joint evolution of two feeling-related behaviors—crying and the response to crying—may be occurring across members of our species. This would be a code, an internal language of shared importance to both the individual and the group, but still gameable like anything in biology. Deception is always lucrative up to a point, but if the deception is rare enough, the whole program of crying and responding could retain value as a truth channel.

For our species, as well as for individuals, such a channel could have been favorable once we became complex socially cognitive beings, capable of deception and denial and with strong voluntary control over our expressions—since if all emotional readouts can be gamed, then all mean little, and social communication loses a great deal of value. Thus an arms race between truth and deception ensues: pausing when cognitive control (benefiting the individual who can achieve this control) is finally attained over the new signal (which then loses its truth property of some value to the species), and restarting a million years later when a misguided axon pathway blunders unexpectedly into a new patch of cells in the brain, perhaps those governing surface physiology of the skin—resulting in blushing, crying, and whatever comes next.

Since emotional crying is evenly distributed across humanity, we can be certain we did not acquire this trait from Neanderthals, who lent their genomes chiefly to the Eurasian lineages. It is not known whether Neanderthals shared this trait as well—most likely they would have, if crying capability had emerged in a common ancestor that every human being shares with them. The Neanderthals had stable social communities, preserved their cultural traditions, took the time to paint symbolic art even as they were dying out, and buried their lovely children. In my imagination at least, they shed tears like ours, until the end.

•

Mateo was not suicidal, but could be diagnosed with major depression. I attached the label to him that night. Though it seemed an oversimplification, among other defining symptoms of depression he had a prominent hopelessness, expressed as an inability to look forward in time. Without hope for the future, Mateo could only look back.

He never did cry for his family that night—not that I saw, nor that he could ever tell me. In considering this, and the reasons we have for crying, it seemed to me that an odd unity links tears of sadness, when they happen, and the more mysterious tears of joy. Tears come when we feel hope and frailty together, as one. I managed to keep myself from writing this in the chart—or writing that Mateo had no hope left to cry for.

Mild improvements in material outcome that do not require a new model of self and circumstance—as with just making a bit more money in accord with known probabilities of the world—will not cause most people to cry. But when we do cry for joy—as when we feel the sudden warmth and hope of human connection at a wedding, or when we see an unexpected depth of empathy in a young child—there may be a flicker of hope for the future of the community, for humanity, against the cold. We can cry at a wedding or a birth, seeing heartfelt aspiration but knowing deeply the fragility of life and love: I hope that the joy I see here will never die, I hope that the world will be kind enough to let this last forever, I hope that these feelings will survive—but I know very well they may not.

This seems to be a kind of anxiety, even for what we think of as tears of joy, since a threat—though not immediately present—is known, and felt.

At the other, truly negative, pole of value, tears of sadness in adult human beings similarly come not with mild losses from known risks but with sudden adverse personal realizations that must be addressed— like a shock of betrayal when the hope we had for the future is shaken and our model of the world, our map of possible paths in life (a map *is* hope), must be redrawn. When we cry, even when the feeling is negative, hope may be present—with new conditions, but it is hope nonetheless. Then we truthfully, involuntarily, signal this fragility of our future, and the fact that our model is changing, at the moment of realization—we signal this to our species, our community, our family, and ourselves.

Does evolution really care about hope? As abstract as it seems, hope is a commodity that must be regulated in careful ways by living beings— metered out in quantities only enough to motivate reasonable actions. Hope when unreasonable can be harmful, even deadly. Every organism must ask in its own way: when to struggle on, and when to save energy and reduce risk by waiting out the storm? Rage or rest, fight or hibernate, cry or not—all life needs to make such choices, to compute the harshness of the present world—and if the challenge cannot be overcome, to withdraw from the fray. The circuitry of hope control needs to work, and work well. With the high heat of our primate lifestyle—a quarter of our calories are burned by the brain alone—the ancient circuitry of withdrawing from actions may extend in our lineage to the giving up of hope itself, of a sometimes costly conceit occupying our brains rather than our muscles.

Ancient and conserved circuits were already available to help our evolution build this capability—even cold-blooded fish can make the choice to meet adversity with passivity rather than with action. In 2019 cells were studied across the entire brain of the tiny zebrafish (related to us as a fellow vertebrate, with a backbone and much the same basic brain plan, but small enough and transparent enough to let us see all the way through, using light to access most of its cells during behavior). Two deep structures in the fish brain, called the habenula and the raphé, were observed to work together in guiding this transition from active to passive coping with a challenge (the passive coping state is where the fish will no longer try to expend effort to meet a challenge).

Neuronal activity in the habenula (provided by optogenetics) was discovered to favor passive coping (essentially not moving during a challenge); in contrast, activity in the raphé (source of most of the neurochemical called serotonin in the brain) favored active coping (vigorous engagement with the problem). By optogenetically stimulating or inhibiting the habenula, it was possible to instantaneously turn down or up the simple likelihood of the fish expending energy to meet a challenge—and when the raphé was optogenetically controlled instead, the coping effects observed were opposite to those seen with habenula manipulation.

Years earlier, optogenetics and other methods had implicated these same two structures in mammals, in the same basic kind of behavioral state transitions, and with the same directionality of effect in each structure. Now seeing these results emerging in the distantly related zebrafish, it could be said with confidence that the biological foundation of suppressing action, when good outcomes are nearly impossible, is ancient, and conserved, and powerful—and thus likely to be important for survival.

Any small animal can find a crevice or a burrow and stop moving to cope passively with adversity. Even the tiny nematode worm *Caenorhabditis elegans* appears to calculate the relative value of actively foraging or remaining in place, with the full power of its 302 neurons. But larger brains contemplate many more possible actions and outcomes, ruminating and worrying, mapping out decision trees thickly ramified with possibilities projecting far into the future. A passivity of thought is also needed, perhaps—a deep discounting of the value of one's actions, and also of one's own thoughts. Hope draws resources from our attentional and emotional budgets, and perhaps it is best to save the striving and the struggle, and to spare the trouble of tears when hope is gone.

•

That night in the ER, I had a hard time figuring out how to help Mateo. The hospital was busy, and there was no room available for him. As he was not suicidal and did not want to be in the hospital, I could not easily admit him to our locked ward, but our open ward was full. There were transfer possibilities to other hospitals, though after talking it all over with Mateo and his brothers, we ended up sending him home with them—and with an appointment for outpatient care, therapy, and medication—but not before I took the time to carry out an hour of predawn psychotherapy with him, right there in the ER, laying groundwork.

When we can, we often steal the time to do this in psychiatry, almost instinctively, even during the besieged rush of an on-call shift, even in cramped and awkward confines like Room Eight that night. It can be hard to hold us back, as hard as it is to hold back surgeons from cutting to heal; we all live, and move, in the crafts we have built for ourselves.

Without the right foundation, nothing works in psychiatry. Without structural threads to weave upon, there is no new pattern that can be created. As psychiatrists, our first instinct is to start to link together what recovery will mean for that person—the intertwined threads of the biological, the social, and the psychological—not in a rush, but in awareness of the time that will be needed to construct something strong and stable. We do this even if we may never see the patient again, as I suspected that night; I was discharging Mateo to the care of his family, and to outpatient treatment. I would continue rotating on through the hospital, in my own ecliptic path, while Mateo would follow his arc in the universe; in all likelihood our paths would never intersect again.

But the amount of time I was taking was extreme, I realized after nearly an hour had elapsed. It was not until the call shift ended, and I was driving home with tears leaking from my eyes, diffracting the traffic lights, that I saw a larger picture—and saw that it was also about another human being, another patient.

I took so long with Mateo that night because I had been unready for him, for that particular hell, as I had been only once before—and so the therapy was for myself as well, for my own tears that were coming. A connection across time had been formed in my mind. It was only with those tears that I saw the link with Andi, who had brought me to the same place and for whom I had been just as unready. Andi, the little girl from years ago with the brainstem finding—long gone, on a journey none could share.

This time, I had thought I could do something—not much, but something. And that matters—realizing at a place and moment you have been called to be whatever it is that humanity can be for a person. That is not nothing.

•

Years later, following our optogenetics and BNST anxiety work, an even deeper connection between Andi and Mateo revealed itself. There was a curious commonality to these patients, who represented the two lowest moments that medicine had brought me to, from which I had to work hardest to emerge. What had actually brought each of them to the hospital the night I was on service had been fibers failing in virtually the same deep spot of the nervous system. This spot was the base and bedrock of the brain, in the pons, where eye movements and tears and breathing are controlled, and where next-door neighbors in my patients were disrupted—the fine chords, the sixth and seventh, of lost harmony.

But the significance of this, if any, I cannot define. I know only that the site is deep, and old.

The naturalist Loren Eiseley wrote that a symbol "once defined, ceases to satisfy the human need for symbols." Eiseley collected observations from the natural world and recorded the ideas stirred in him by these images as symbols—like a spider out of season, surviving in the dead of winter, having built a web by an artificial source of heat, the globe of an outside lamp. He was moved by this image, despite his near certainty that "her adventure against the great blind forces of winter, her seizure of this warming globe of light, would come to nothing and was hopeless. . . . Here was something that ought to be passed on to those who will fight our final freezing battle with the void . . . *in the days of the frost seek a minor sun.*" Hope, represented by complex life fighting on in the face of inescapable cold, moved Eiseley, and moves scientists and artists similarly. It is close to the heart of what moves us to tears.

For Mateo, there was no hope left to cry for, now that his wife and baby were gone. His lack of tears was also his blindness of the future. But, in some form, I knew, or thought I knew, that he could love again, in time. Hope was not dead, though he could not see this, and so the tears came for me, and not Mateo.

The true end of hope shows up only as extinction, when the last member of a sentient species eases to the mud alone. In the history of our lineage, this finality would have become real many times over, in the fine lost branches of our greater family tree. The Neanderthals and others, in the last traces of their last days, lived out that tragedy for which all else is metaphor.

Extinction is normal. Each mammalian species, on average, gets a run of about a million years, it seems — with a few close calls thrown in the mix until it finally happens. So far, modern humans have lasted only about one-fifth of this interval — but already we have survived some mysterious crises that can be inferred from human genomes, when effective breeding population sizes around the world may have plummeted to a few thousand individuals.

Such demographic events alone could help explain the prevalence of odd traits with little obvious value — behaviors somewhat unrefined, having found (like crying) incomplete purchase in the population due to only a mild benefit. When a species goes through a population-size bottleneck — where only a small fraction survives or migrates — whatever traits were present in the chance survivors (or migrants) then for some time enjoy outsize prevalence, whether or not the traits were unusually important to survival. This could be the case for crying emotional tears—and help explain the seeming uniqueness of such a trait, among animals.

On the other hand, perhaps we needed this truth channel more than other related species—in building ever larger and more complex social structures over time. Crying could have come along as first just a misrouted brainstem projection, but the responsible genetic variant might have found purchase in the mingling populations of East Africa as our modern lineage arose, when we used our fingers and brains to build one another houses, constructing durable communities at great cost. Perhaps tears were needed after we had grown too good at the last forgery, at gaming the last signal of grimacing or keening. Builders need solid ground; social builders need ground truth.

The last Neanderthal—a big-brained, bruising, nearly modern human, the last member of a branch of our family tree that buried the lost with ritual and care—died just an eyeblink ago, clinging until the end in caves near what would become the Gibraltar seashore, in final retreats hidden from, as Eiseley said, "the first bowmen, the great artists, the terrible creatures of his blood who were never still." They may have cried at weddings, at births—but when the last starving Neanderthal watched the last baby trying desperately to nurse, skin on skin but fluid failing in the ducts . . . there was no hope left for doubt, there was no future left to question, or to fear. There were no tears then, under the moon without answers—just a dry streambed, set back from a salt sea.

CHAPTER 2

FIRST BREAK

Horns of the long-lived stag began to sprout,
The neck stretched out, the ears were long and pointed,
The arms were legs, the hands were feet, the skin
A dappled hide, and the hunter's heart was fearful.
Away in flight he goes, and, going, marvels
At his own speed, and finally sees, reflected,
His features in a quiet pool. "Alas!"
He tries to say, but has no words. He groans,
The only speech he has, and the tears run down
Cheeks that are not his own. There is one thing only
Left him, his former mind. What should he do?
Where should he go—back to the royal palace
Or find some place of refuge in the forest?
Fear argues against one, and shame the other.

And while he hesitates; he sees his hounds,
Blackfoot, Trailchaser, Hungry, Hurricane,
Gazelle and Mountain-Ranger, Spot and Sylvan,
Swift Wingfoot, Glen, wolf-sired, and the bitch Harpy
With her two pups, half-grown, ranging beside her,
Tigress, another bitch, Hunter, and Lanky,
Chop-jaws, and Soot, and Wolf, with the white marking
On his black muzzle, Mountaineer, and Power,
The Killer, Whirlwind, Whitey, Blackskin, Grabber,
And others it would take too long to mention,
Arcadian hounds, and Cretan-bred, and Spartan.

The whole pack, with the lust of blood upon them,
Come baying over cliffs and crags and ledges
Where no trail runs: Actaeon, once pursuer
Over this very ground, is now pursued,
Fleeing his old companions. He would cry
"I am Actaeon: recognize your master!"

But the words fail, and nobody could hear him.

—From Ovid, "The Story of Actaeon," *Metamorphoses*, Book III

An image can take root and grow. Here, it is of a young father with his two-year-old daughter in the 767, slowly banking harborward, nearing the burning steel tower—it's a frame of the moment when he at last knows the impossible truth, his pulse thudding thickly but she's calm amid the chaos, because Daddy said there were no monsters. He's turned his daughter's head firmly toward his own—she's a frail warm spot glowing within an infinitude of cold—for a moment of silent communion before their sublimation.

A little girl and her father, searching each other for grace as the plane roars into the second tower—this wordless image became physical, sown across the world into the arable mind of a man named Alexander, as he sailed through the Cyclades. Quickened, germinating, the imagined scene gathered form—investing all the soil of his thoughts, insatiably drawing to itself all the fluid of his soul.

•

The fundamental rules of Alexander's life had already been rewritten just before September came, and so it may be that his brain—fallow for decades—was ready when the outside world was transformed as well. In 2001, as the shortening days of late summer brought chilly afternoons and crimson leaves to the San Francisco peninsula, Alexander stepped down at sixty-seven from the insurance company where he had labored for decades—as a fairly effective underchief, but where he was no longer nimble enough for the shifting strata of Silicon Valley. His domain would now be at home only, among the coastal redwoods of Pacifica, in the high-raftered house he and his wife had built in a foggy ravine twenty years earlier—big enough for their three sons and perhaps some grandchildren. He was a stately man, slightly bent, in the growing calm.

No warning notes had sounded in his life, no explanatory story was found that his family could share, by the time I met them in the emergency room six weeks after September 11. By then, his whole world had been blasted apart—not by exploding jet fuel, but by ferocious, exuberant, unstoppable mania, bearing no resemblance to anything that had come before in their lives. It was first break—that moment when links to reality snap in response to a windstorm of stress, or to the scythe of trauma, or to other triggers unknown—and the human being first comes untethered. First break, when those with mania or schizophrenia are cut loose—at great peril—and sent aloft by their disease.

In September, when the storm tide began to rise, Alexander had been only marking retirement—sailing the Aegean with his wife, traveling in antiquity. Now less than two months later, back home, he was transformed, brought to my emergency room by police and family. What had been swept along by the hospital process and settled, what I saw first, had no visible flaw. Not knowing him, I saw only a crisply alert man scanning the newspaper with intensity, fiercely cross-legged next to his gurney.

The elusive, protean mystery of psychiatry came next—discovering what it was that had changed for this person, and why. No brain scans exist to guide diagnosis. We can use rating scales to quantify symptoms, but even those numbers are just words transformed. So we assemble words; this is what we have. Phrases are pulled together, and molded into a narrative.

The people involved talked, all of us—in different combinations: the patient, the police in the hall, the family in the waiting room—all searching for the right frame. For someone with no mania in his past or family: why him, why now? He had experienced the day itself, the strike to the heart of his country, no more intensely than any other person.

Even the pain he had felt, in empathy for the lost, by itself did not merit this extraordinary consequence. Death comes badly for conscious beings and always has. The unthinkable is universal, but mania is not common. Nevertheless it came for Alexander—after a delay.

For a week after September 11, Alexander was just a bit on the stoic side, only revoicing the common thoughts of shock and pain around him. He read stories of the victims, but then began to focus on two of them, a father and a daughter, a pairing he had not lived personally. A scene emerged and grew more detailed, and he spoke to his family of imagining their final moments—while inside his brain, a secret remapping had begun. In ways still mysterious, new synapses were formed, and older connections were pruned away. Electrical patterns shifted, as scripts were overwritten. For a week his biology silently learned its new tongue, and then it reached out, expressive at last.

The first manifestations were physical. He nearly stopped sleeping, becoming fully alert and charged with life for twenty-two hours a day. Never the chattiest before, Alexander now could not hold back a vast volume of words that came out in a pressured torrent—turbulent and interjaculating—yet still coherent, at first. The content of his speech changed too—he was saltier, charismatic, uplifting, and illuminating. Beyond language, his whole body was affected; now ablaze in new youth, he was suddenly voracious and hypersexual. No old bull out to pasture, he was an organic being newly ready to react, to interface—his skin surfaces were functionalized and available. Life was lacier, compelling, alluring.

Projects and goals came next. They were valorous and numerous, with a tinge of excitement, a subthrill of risk. He bought a new Dodge Ram pickup, with a heavy-duty trailer hitch and extended cab. He ran all night, read books all day, and studied theory of war, writing pages on the movements of forces and reserves. A theme of self-sacrifice appeared and grew stronger; he wrote letters volunteering to join the navy, and was found one evening rappelling down a redwood trunk in the fog, training for war. He was breaking from his lifelong chrysalis, transforming into a newly emerged monarch.

There had been a certain charm to the transition, up to a point, but then he veered into thoughts of good, evil, death, and redemption. He had dwelt until this event in a sort of unperforated Lutheranism, stormless and modestly nourishing—with minimal connectivity to any other part of this life. Now he began to speak with God—first calmly, then frantically, then in a scream. Between these prayers there were sermons for others, in which he became irritable—swinging between euphoria and crying.

Nearing midnight before admission, he ran from the house with his quail gun, throwing branches and bark at his sons as they tried to stop him in the yard. Police found him two hours later, at bay in a thicket near a dry streambed, ready to strafe the stinkweed. They took him and subdued him with the mundanities of medical-legal incantations, all the energy still brimming up behind his eyes like tears.

The fury had dimmed externally over the next several hours in the hospital. By the time I spoke with him, he had only a rhythmic motor pattern, like a caged pacing lion, except it was a vocalization, a refrain again and again: *I just don't understand*. With clarity, with surety in his own form and role, he could not understand his family's reaction—why his every action did not seem perfectly logical to them, an example which all should follow.

The fixity was striking, and pure. Alexander's first break had been a clean separation, without the messy compound breaks of psychosis or drugs. He was dislocated. Unchurched.

What next for the new warrior—dopamine-receptor antagonists perhaps? He did not want help, saw no need for our process, and refused treatment. In the closed system of his pressured logic, there was pure clarity, and explosive danger. An irresolute messenger, I wavered before him as he described the image that had grown in his mind, of the girl on the plane, with the father holding her head gently and firmly, locking eyes so she could see only him until the end.

Pictures came to me, intense associations. The unique abstraction of psychiatry—science with language, medicine with text, upon which the most effective care is built—allowed me to spend each day immersed in words and images, moving beyond story to allegory, even if doing so was futile—in dialogue with history, with neuroscience, with art, and with my own experience. Here the first story that came to me, provoked by his transformation, perhaps prompted by the image of Alexander sailing among the Greek isles, was Ovid's hunter Actaeon—son of a herdsman—who was turned into a stag as a punishment from the angry goddess Artemis, after he was caught spying upon her bathing in a stream. He had new strength, new speed, a new form—he was given strong horns and fleet hooves—but the timing was off, the context was all wrong. He had become a prey animal in the midst of his own hounds—Blackfoot, Trailchaser, Hungry, Hurricane—who tore him to pieces. Perhaps it was an Actaeon I saw before me, transformed by the moon goddess, with the police and me as Arcadian hounds, and

Cretan-bred, and Spartan—the whole pack, with the lust of blood upon us, baying over cliffs and crags and ledges where no trail runs.

But then . . . unlike the case with Actaeon, whose new form of a stag had no value, for Alexander there was a use, grim and suitable, for the new form he had been given. In this sacrifice, he was perhaps more of a Joan of Arc—like Alexander, born far from military life. In her case it was on a small farm in Lorraine where the mystery began to speak to her. Without trying to diagnose a historical figure—always tempting but usually unwise for psychiatrists—I could not help but imagine how her alteration happened to briefly work well for her. When she was only seventeen, as France began to fall before English armies, she emerged into a new way of being—not disorganized as in schizophrenia, but goal-directed, focused on continental politics and military strategy. She talked herself to the side of the Dauphin with a firm conviction that she was essential, and with a powerful religiosity that allowed her to infuse the fight with a spirit seen as divine—bearing banners not swords, living through maelstroms of crossbow bolts, advancing through her own blood to the coronation.

Alexander's transformation also arose in the pastoral quiet of a country in peril—an alteration created by that very peril—and his new form fit the crisis. Some details were imperfect—the tenor of current culture was mismatched for what he had become—and he was the wrong vessel, but then: was he any less-suited than a seventeen-year-old peasant with no training in tactics or politics? By the time she was captured by the English and burned at the stake, Joan had already saved her country and won the war. And here we were about to cure Alexander, to cauterize this illness, to burn away the spirit. A bumpkin psychiatrist, I stood ready, with my medieval tools.

And there, in that uncertain moment—the two of us caught in a tiny personal crosscurrent, lost in a vast global atmosphere that had been marred for months by burning flesh and the slipstreams of airborne predators—a thin and fragile tendril of memory, of my own story, swirled to the surface.

.

I was leaning against a chain-link fence, the perimeter for an outdoor platform of the T—Boston's subway. It was near midnight on a chilly October evening. Drained after a long day in the lab and a failed experiment, I was fatigued and irritated. The area was nearly deserted, except for two men talking calmly at the other end of the dimly lit platform. A pair of silhouettes, one tall, one short. For a minute of peace I closed my eyes and bowed my head as we waited together.

When my eyes opened to look for the train, I saw an eight-inch blade, silver and gold in the subway glow, fine at the tip, even tender as it nearly touched my shirt, nearly part of me. I saw only the beautiful blade, in incredible detail, and everything else was blacked out; I was swept in, there was nothing else in the world, and in that moment, I felt an awareness of all the events and interactions and steps with which the world had positioned me here, and seemed to understand that this fate had been prepared for me with care and affection. I had come to where I should be, and an odd peace, a grace, came over me.

I surrendered my backpack, waiting passively as it was emptied by the tall shadow, keeping my eyes focused firmly on the blade held by the other. My sweet misericord, the thin blade of mercy used in the aftermath of medieval battles, dispatching the dying at Orléans and Agincourt. The steel seemed to pulse in the surreal light of the subway platform, with every cell of my body locked into its rhythm.

The backpack contents were exposed—which I knew to be only a developmental biology journal and seventy-five cents for the subway fare—and the next memories are fragmented. A burst of angry words, the knife seemed to twitch with unclear intent, and then I was suddenly no longer passive. I remember sweeping my left arm up and out, creating a narrow space to break away to the right. In my next conscious memory I am blocks away, not knowing where, running alone through the star-cold night.

There followed in me a high energy over the next few weeks, a bubbling up inside of anger and euphoria, a feeling in my chest like a geyser preparing to erupt. Then the sensation eased to a week or two of light pressure; then everything was distilled down to a calm clarity—

and finally . . . nothing. It was gone, never to return—a minor deviation, a ride, a day trip, real but weak in me, never breaking through.

As I considered Alexander, it seemed his brain, unlike my own, must have been prepared—soil truly fallow, fertile and waiting for the seed. Yet even he might have escaped mania if not for September 11; mania comes at great cost, and his brain had set a high threshold, tuned to only respond this way to a seemingly existential threat to the group: his whole community appearing at risk, with invaders descending. His stolid odyssey of the useful and the good ended only with burning towers, and his transition when initiated was swift and sure, a second puberty, remapping him a final time. Steroid stress hormones coursed through his brain like juvenile hormone through a caterpillar, sweeping away the squirming and helpless peacetime stage, the old larval neurons killing themselves, involuting—ruthless, precise, meticulous. On into mania—wings for the mind. Metamorphosis.

Perhaps I lacked the genes, the temperament, the mental landscape to fully quicken. Or perhaps it was that my circumstances had been different from Alexander's; I had been alone, the assault had been directed only at me, not my community—and I could run, needing only two minutes of pitch-perfect adrenaline-related neurochemicals to meet the threat, just that elegantly tuned fight-or-flight response. A stable behavioral change, lasting weeks or months, would have made no sense. Mania, at least in those cases when symptoms and threat align (as they did for Alexander) would seem more of a durable and social fury, by some design or chance prolonged for defense of the community with goal-directed activity, but only if a new way of being were needed, an elevated state. Mood elevation has the capability to bring forth energy for social construction—for the time needed to build defensive earthworks under the rumor of war, to migrate the drought-stricken clan for weeks toward water without sleeping, to harvest all the winter wheat when the locusts hatch—and with all the rush of a positive charge, that rewarding feeling needed to upend preexisting priorities temporarily, to align a person's whole internal value system to meet the crisis.

But in our world, mania is fraught with danger: harmful to the patient and costly to the community; it is the exception rather than the rule for symptoms to even appear appropriate. Thwarted in the modern milieu by our intricate conventions and rigid rules, the incompletely hatched monarch is trapped in a cracked and hardened casing—with new wings caught, beginning to rip in the raging of the struggle to emerge.

As we talked, I could feel the room charged with this trapped energy. In his irritation and agitation, Alexander unknowingly was spawning imagined scenes of his life in my mind that took root as the airplane scene had for him, unspoken but oddly clear and detailed. I let the vision grow, and saw his own eyes opening in his living room, back home from his odyssey in October, to a castrated dog on the rug with its bloated belly irritatingly exposed, breathing stertorously out of step with the Pachelbel coming over the dusty stereo. The dog was Alexander of the last thirty years: weak, infertile, asynchronous. The need to leap up and strike out—to act—would have surged.

A hiking trip to a coastal estuary was proposed by his wife, to spend some quiet time among the elegance of the local herons, but what mattered to Alexander actually were the shrikes and butcher-birds of the desert, the Predators flying over Mazār-e Sharīf. Called to serve, Kandahar's time had come again—to march from Macedon to the east one more time. He would have felt a rising swirl of rage. No, it was of libido. His ducts would have felt filled with fluid, all of them, with the ductal smooth muscle tensing against what he had stored within for decades. Squeezing on what he had, what he had to give. Strong as jet fuel.

•

There had been no natural way of stopping the birth of this new being, any more than stopping childbirth, and mania can last for weeks or more on its own. But in the hospital, any parturition can be slowed or stopped, for a time. When Alexander demanded to leave the ER, triggering frantic pleas from his family, under my care his freedom was taken and civil rights temporarily stripped away. Thus roped to the mast, he received olanzapine—dopamine and serotonin modulation, to block mania's siren song—and within a week he was, as we say, normalizing.

But there was an unapplausive feel to the outcome; normalizing him was not a clear-cut win. The clinical team exchanged no pleased commentary on rounds. Instead there were only hesitant and fragmented conversations in the housestaff room on the meaning of mania, and on the ethics of intervention.

Mania cannot be trivialized or romanticized. As interesting as the state is—and as euphoric as the patients can be, at least briefly uplifting all around them with their contagious belief in what might be possible—mania is destructive. In the vulnerable, those predisposed to bipolar disorder, mania is often not threat-triggered at all, and does not even approach utility; rather, it is unpredictable and can be accompanied by psychosis, a breakdown in the process of thinking, the catastrophe of suicidal depression, and death.

Any value of mania today is inconsistent, but increased-energy states are consistent: a common heritage of humanity across cultures and continents. Not all of these states fit exactly into the same frame. Variants might include *amok* in Malaysia, a state of intense brooding followed by persecutory ideas and a frenzy of activity, or *bouffée délirante* in West Africa and Haiti, a state of sudden agitated behavior, excitement, and paranoia. Both of these, and mania itself which is seen around the world, may represent thin slices through a much broader and complex multidimensional structure, a cluster of possible behaviors, of altered states. Different cultures take their own cross sections to describe these states, each at a distinctive angle.

Human evolution clearly has not converged upon a single or ideal strategy for sustained mood elevation, if there can be one, and many different genes are linked to bipolar disorder. Telling the stories of past struggles in human evolution, our genomes are laden with other first-pass fixes that still need refinement. Across much of modern medicine outside of psychiatry, it has long been possible to ask, and even answer, why a genetic disease might be common. To explain the persistence of the blood disease sickle-cell anemia, for example, we can tell stories of our coexistence with the microbial parasite *Plasmodium malariae*, which evolved along with us, driving adaptation of our blood cells and immune systems in an agonizing call-and-response playing out over millions of years.

Sickle-cell disease and the related diseases called thalassemias (a classical name, for their Mediterranean distribution) are burdens borne by many modern human beings who share peri-equatorial genetic roots, where *Plasmodium* and its mosquito carrier thrive. The burden takes the form of mutations in hemoglobin, the protein in our red blood cells that delivers oxygen to mitochondria (like *Plasmodium* once immigrant microbes, our mitochondria are now fully symbiotic partners in our survival). *Plasmodium* lives in our red blood cells if it can get in, and these mutations in hemoglobin work against this ancient enemy, suppressing malaria as they block the spread of *Plasmodium* through the blood. The mutations also, however, bring the risk of misshapen red blood cells that cause disease symptoms: pain, infection, and stroke.

As in cystic fibrosis, the human carriers with one mutant gene usually have no symptoms, and it is only when two mutant genes come together that the sickle-cell disease state is created. But unlike those who carry a single cystic fibrosis gene (at least as far as we understand it today), sickle-cell carriers (the nondiseased, carriers of only one mutation) have a clear benefit, resisting malaria and thus revealing a harsh evolutionary bargain: a steep price is paid only by those with two copies of the mutation for a benefit enjoyed by others with one copy, who do not suffer. Thus these mutations are ragged measures, quick hacks, still jousting in the tortuously slow arena of natural selection.

The lesson of the sickle cell is that the disease, and the diseased, make sense together only in the broader frame of the human family and its evolution. Though it is not always easy for scientists to find these perspectives, the simple fact of achieving an explanation has been important, helping to liberate us from the grip of mysticism and blame. Psychiatry, however, has continued without such insight. More consequential than any other type of disease in terms of the immensity of death, disability, and suffering caused around the world, mental illnesses nevertheless have remained essentially unexplained in this way, and no definitive explanation is possible now.

Yet neuroscience has reached a tipping point. For the first time, scientific explanations for what these illnesses are, biologically, seem within reach—and as with sickle cell, as with all of human health and disease, the prevalence of mental illness should be informed by evolutionary considerations; as Theodosius Dobzhansky wrote in 1973, nothing in biology makes sense except in the light of evolution.

But thinking about survival and reproduction trade-offs can be misleading if the questions asked are naïve or incomplete. For example, the harm of psychiatric disease to patients seems clear, but who would be the recipient of the evolutionary benefit, if any, that allows these traits to persist? With the sickle-cell trait, those who receive the benefit are not the same as those who suffer. Is this also the case for mental illness, that there is some benefit only for close relatives? Or alternatively, could it be that the mentally ill do directly benefit—at some time, in some way?

We must acknowledge that the present-day world could provide no answer—evolution is slow, while cultural changes are fast, and society is not close to a steady state. We are likely to be imperfectly suited for our world as a result. But there is hope for understanding; those traits and states we do have likely mattered for survival until very recently, if not right up to the present. What doesn't matter for survival quickly disappears, leaving only traces, footprints in the damp sand of genomes, fading with the waves of generations. In the lineage of mammals, egg yolk genes were lost as soon as milk evolved (though broken fragments of the yolk genes persist, even within our own genomes still). Cave fish and cave salamanders—in sunless colonies, blocked off from the surface world—lose their eyes after generations of darkness, leaving skin stretched over sockets in the skull, relics of a sense no longer needed.

To understand this oddity of its design, a cave salamander would need to know about something beyond its conceptual reach—the illuminated world of its ancestors, and so the value of the twin holes in its skull: pathways for information in an ancient world of light, but only vulnerabilities in the modern world. Likewise inexplicable depths of our own feelings, and our own weaknesses, might also be best understood in the context of the long march to our current forms, finding little in the present world explanatory. But caution is needed: not only do we lack data, but also our imagination itself is subjective, and our perspective is limited and biased. The borders separating broken and unbroken may shift, and blur, and even fade away as we approach.

Currently it is impossible to be definitive about the role of evolution in mental illness. But human origins and evolution have to be part of the picture in thinking about psychiatry—as with anything in biology reflecting conflicts and compromises that arose, and were tested, over many generations. A pure hunter-gatherer, more than one hundred thousand years ago, might not have needed the prolonged intensity of mania, and would have benefited from simply cutting losses and moving on from threat or conflict—to new vistas beyond the horizon. But when we build—as we have more recently, in the form of houses, farms, communities, multigenerational families, culture—existential threat might be best met with an elevated state of being, even if unsustainable.

Neuroscience has made little progress in understanding mania, or the bipolar disorder syndromes, which form a spectrum of severity all sharing a manic-like state. Indeed, mania is not binary but ranges in degree from mild hypomania (a sustainable mood-elevated state not needing hospitalization) to recurrent spontaneous manias (increasingly severe with each episode—psychotic even, with breaks in perception of reality—ending in a dementia-like state if untreated).

Neuroscientists interested in mania have explored certain types of brain cells with relevance to the core symptoms. For example, dopamine neurons have attracted attention for their known roles in guiding motivation and reward seeking—elements clearly overabundant in mania, on display in that remarkable symptom we call "increased goal-directed behavior," and exemplified in the numerous projects, investments, plans, and energy of Alexander's rebirth. Circadian rhythm circuits have also been pursued, since one of the most striking features of mania—used in diagnosis as well, and prominent in Alexander—is a profoundly decreased need for sleep. This symptom is especially interesting since mania does not cause poor sleep per se (nor attendant problems that come with the poor sleep of insomnia such as lethargy, somnolence, and the like). In mania there is a true decreased *need* for sleep—as Alexander experienced—with maintained high functioning of brain and body over prolonged periods, and very little rest taken or required.

Are these dopamine and circadian circuits clues, then, embodying pathways into the mystery that is mania? In 2015 the dopamine and circadian aspects were brought together with optogenetics. Mice with a mutation in the circadian rhythm mechanism, in a gene called *Clock*, were found to show behavior that could be interpreted as manic-like, in the form of prolonged phases of extremely high movement levels. This state was found to occur at the same time as phases of higher dopamine neuron activity. Could that dopamine elevation be causal, driving frenzied movement levels in mice? Using optogenetics, the team found that increasing the activity of dopamine neurons could indeed induce the manic-like behavior; furthermore, suppressing dopamine neuron activity could reverse the manic-like state of the *Clock* mutant mice. We are far from a deep understanding of mania, but optogenetics has helped unify two of the leading hypothesized circuit mechanisms. Going forward, it may be useful to consider that the dopamine neuron population is not monolithic but composed of many distinct types that can be separably identified early in mammalian brain development; future work may allow targeting specific subtypes relevant to mania, such as those specific dopamine neurons that project to brain regions involved in generating actions and action plans.

What other genes relevant to mania are found in human beings? Bipolar disorders are heritable, running strongly in families, but there are few single genes that, by themselves, can determine the disorder— in fact, there may be dozens of genes or more that each contribute small effects, as with height. A few of these genes have turned up fairly consistently with scanning of whole human genomes in studies of type I bipolar—the kind with spontaneous and severe manias, among the most strongly heritable of psychiatric diseases. ANK3 is one such gene, which directs the production of a protein called ankyrin 3 (also known as ankyrin G), which organizes the electrical infrastructure of the initial segment of the axon—the first section of each threadlike output wire, that electrical-information transmission line connecting each brain cell with all of its receivers across the brain.

These mutations, which contribute to causing bipolar disorder in some people, likely lead to insufficient ankyrin 3 production. In 2017 a mouse line was created with "knocked out"—insufficient—ankyrin 3. These initial segments of the axon were indeed poorly organized in the knockout mice as a result, in an interesting way. Inhibitory synapses that would normally be clustered at that crucial spot of the axon, acting like dampers to prevent overexcitation, were gone. And the mice showed some manic-like properties: much higher physical activity levels, in terms of both general locomotion and movements specifically directed toward overcoming stressful challenges—that is, elevated goal-directed behavior. Amazingly, this pattern could be blocked in the mice by medications, including lithium, that are highly effective for bipolar in human beings.

As interesting as ANK3 is to psychiatrists and neuroscientists, in human beings its mutations cannot alone explain all mania, and bipolar disorder in general is far from understood. We also do not understand the association of mania with depression—that other "pole" of bipolar. Manias frequently end in deep depression, and many patients cycle from up states to down states: mania to depression, or depression to hypomania and back—but nobody knows why, and the ANK3 studies do not offer an answer. Is there a neural resource of some kind that is consumed by mania, leading to a slide into depression? Or perhaps instead there is overcorrection from a system responsible for turning off mania when the threat is past, but occasionally overshooting? An imprecise hack indeed—one that in the past could be tolerated by our species as a whole, if not so much by those living with it.

•

The evolution of civilization is far faster than the evolution of biology. The global reach and power of individuals, across space and time, now make hypomania and mania more dangerous, and more destructive.

Certain historically significant figures have, like Alexander, no doubt borne this burden or one like it, all trying to meet the challenges of their time—and for a brief moment have found themselves moving toward a state of vigor, optimism, and charisma that from some perspective is an elevated expression of what a human being can be. But disaster would have come for many of them—and for Alexander, born in the wrong time and place, there was no safe opportunity to complete the metamorphosis, to fulfill that calling.

Upon discharge from the altered world of the hospital, as when leaving Baum's land of Oz, every patient seems to receive a parting gift in some form. On surgical services, some patients even receive a new heart. As we say in the hospital, in psychiatry, most patients are Dorothies—they just get to go home. This was the only path for Alexander: coerced treatment, renormalization, and release back to his community—the aligned goal of everyone who cared for him.

At follow-up one year later, Alexander's wife described him as "better than ever." The shadow of his illness had been that darkness shining in brightness of Joyce's *Ulysses*; this was a darkness that the brightness could not comprehend. Though no longer manic, he still could not bring himself to repudiate the state he had entered into, nor his actions in the state. He still didn't understand why we had acted as we did. I thought him a little glum about it all, but in the end, he had been given a way to live with his wife again, to slip into retirement without redirection or consequence, and to hike in the heronry.

CHAPTER 3

CARRYING CAPACITY

Tonally the individual voice is a dialect; it shapes its own accent, its own vocabulary and melody in defiance of an imperial concept of language, the language of Ozymandias, libraries and dictionaries, law courts and critics, and churches, universities, political dogma, the diction of institutions.

—DEREK WALCOTT, *The Antilles: Fragments of Epic Memory; The Nobel Lecture* (1992)

"I had a teratoma in Paris," Aynur said. "It started from an egg in my ovary, and sprouted teeth and living neurons and clumps of hair, all twisted together, growing in my belly. The French doctors took the tumor out, but after the surgery it was hard for me to walk, or bend, or sit upright. I was living alone, and I had to do everything very slowly.

"I was in this state, when I got a strange letter—with twelve pictures of our hometown, unexplained—in the mail, from my mother. I remember stepping carefully across my loft to spread out the photos on my breakfast table.

"I felt something of the warmth of home. It was like my mother was touching me physically, reaching her hand out across the whole breadth of Europe and Asia. The photos showed familiar street views, buildings crowded against the roads, with our rounded windows and wrought-iron balconies, and with people standing out vividly against the gray skies of fall, like droplets of dye.

"The colors on our clothes—you could never see such a scene in Palo Alto. Deep reds, rich indigos, really brilliant yellows—and all pigments from nature—dark brown from walnut skin, that feathery purple from the tamarisk tree. You might have seen these dyes on our silks, the Uighur kind we call *atlas*—which means graceful silk. It's soft but strong, used for women's costumes and ribbons and wall hangings. The world might know of our silks, I think, if nothing else about us. And there are similar color styles on our everyday work clothes, even on manufactured jackets—bright purple and peach and orange and gold, mass-produced garments that we get trucked from the capital in Ürümqi—it's all the same spirit, just our taste, the contrast of strong colors.

"But there was a problem. The longer I looked at the photos, and the note, the more I felt there was something strange going on. There was no explanation in my mother's short letter, no comment on the pictures themselves—and her actual written words were just a dry response to my own last note.

"I had emailed her a long update on my graduate work—and since there had been no news from my husband for two weeks, I had asked her if I should come home for a visit. I looked back at my mother's note, and reread her words: 'You should not come. It is still too hot here, and now you are not used to it. You have been so long in France, you should stay there.' But actually it was France having heat waves, and I had complained to her already about it. Paris over the summer that year had been hotter than ever, and anyway I could see from the pictures the little boys and girls at home were already in their autumn coats.

"After a few minutes, I noticed something else: there were no young men on the streets. Many children, women, and motor scooters. But all the men, of my husband's age, were gone from the streets. In every photo.

"I remember then in my haste to get out, to find an open Internet café, I almost fell down the narrow stairs leading to the rainy street. A piercing pain from my surgery was just starting to surface as I reached the door to my apartment, but only when I got down to the street did I understand how bad it was. I couldn't go back up. I couldn't even walk.

"There on the street in Paris, it came to me how deeply I was hurt inside. It was dark, and the stones were wet. My family was in danger, and I was alone. And there I found, when I couldn't walk, that I could run."

•

Aynur was animated and exuberant, and broadly smiling at times, in a manner discordant with the darkness of her story and its gathering crescendo of physical and emotional pain. I began wordlessly wondering what natural process in the brain sets the timing for awareness of suffering. At the same time, in a parallel stream of thought, I was secretly awed by her imagery. It was unexpected and effortless, and her story was coming in a torrent that grew rapidly more powerful.

Awareness of anything, including our internal feelings—consciousness, some would say—does not simply flip on and off, as if controlled by a switch. Awareness even of pain gathers itself, seeming to emerge with movement in time, along a trajectory arcing from one moment to another.

Each feeling is intimately entwined with—and perhaps even identical with—a growing and cresting and abating of brain activity. That timescale in one sense spans hundreds of milliseconds, and in another sense, millions of years. Feelings are, like people, paths through time.

The elements of human subjectivity—what we all feel with our conscious minds, and when—may exist in the modern world only to the extent that these feelings caused actions needed for survival before, in the distant past. And so for Aynur and me, with homes at nearly opposite ends of the earth—our feelings in common mattered, and it also probably mattered how they were felt many millennia ago. Recognizing this connection seemed to me a grace of sorts, granted to our long-gone forebears across the cold expanse of time—and also seemed a comfort in the present: acknowledging all the partners in this family conversation, and seeing feelings not as clinical injections of information into our minds from the external world, but rather as connections with each other, across the scattered expanse and long unwritten history of the human family.

At the other end of timescales in biology, just as Aynur experienced the emergence of her visceral pain, our indwelling experiences as individual animals are also defined by movement in time—over a fraction of a second. Each conscious experience is dynamic on this timescale. It manifests, peaks, and lingers—keeping its own pace, separate from the stimulus that brings it forth.

This is a long time for consciousness to take in coalescing—a hundredfold slower than the electrical signaling speed of brain cells in isolation—a matter of two hundred, rather than two, milliseconds. Whenever the world sends us new bits—a pinprick, an unexpected sound, a light touch—almost a quarter of a second passes before that exquisite glow of conscious awareness shines forth. Reflexes are different—unconscious processes can be much faster—but consciousness, for some reason, takes its time.

The individual subjective experience, then—in the moment of our awareness—can be understood from the perspectives of both evolution and neurobiology to not just represent a dump of data from the external world. The tidewaters of the external sensory ocean not only percolate deep but "gather to a greatness," as Gerard Manley Hopkins wrote of the grandeur of the divine—with a mysterious inwreathing through the brain's inland fens and waterways, to manifest finally and fully. Something special is happening.

Neuroscientists have come to know this strange fact of mammalian consciousness from many kinds of experiments, most grounded in direct measurement of electrical activity in the brain. Two or three hundred milliseconds elapse before the response to a ping, an unexpected sound or light, peaks in our cortex—that thin and wrinkled covering of cells wrapped around the brain of every mammal like a shawl.

Not only does this seem like eons of silence to a cellular physiologist like myself (used to thinking in shorter timescales of only two or three milliseconds, across synapses and along axons) but the long delay is surprising to nonscientists as well, to anyone who has observed a cat chase its prey, or a boxer lean back from a jab, or just two people interact in animated conversation—all of which play out on much faster timescales. The trained boxer seems to dodge before it is possible, responding to a specific threat trajectory in less time than it should take if consciousness were required. And human social interaction, especially, seems impossible in light of this number. How clumsily slow it would be, how unlike us, to wait almost a quarter of a second to respond to each new bit of spoken information—or even longer if actual consideration is involved.

And that is only for speech; even more puzzling is the fullness of social interaction, with all its data streams. What of integrating the bit-rich visual inputs conveying eye contact, hand motion, posture? What of each diffident subangulation of a lip, or shift in body orientation—with recognition of every visual stimulus essential for generating an appropriate response? These streams of information need one another to make sense, just as people need one another for meaning. And what of larger interactions: a team or a town hall? Human groups are teeming with conflicting desires, polite or malevolent lies, shifting alignments—each information stream not only playing out simultaneously but involving the others to generate meaning, and requiring constant reinterpretation, and co-interpretation, while the speakers—and their models of the world and one another—all change with time as well.

Deeper insights have license to take longer, and can come much later—after all the information is in, after weeks or months of incubating like a caterpillar in its white-matter cocoon, ensconced in the silk of interwoven axons—until new awareness emerges one day, breaking free fully formed.

•

"For three more months," Aynur told me, "from that moment, I had no contact with my husband. I was so afraid. My parents were afraid too, but very careful. Even when I finally got them on video chat, they said nothing. I did not get to hear if he was alive. If they had heard something, they did not say. I couldn't ask directly about the photos—I didn't know if sending them had been forbidden, or who might be listening. But a wife should be expected to ask about her husband, I think. It would be more strange to not ask. Anyway, it didn't matter what I asked my parents about him—'We don't know,' they said to everything, and that was all.

"Everything was unknown. After two months of this I lost the ability to sleep. It wasn't just not knowing. It was that there was nothing to do about it. I couldn't help my loved ones. I was paralyzed. I was being eaten alive from within—there is nothing like it that you could understand. All was the opposite of your life now, where you control everything.

"Something had crawled into me, and begun to gnaw my spine, and hollow me out from within. There was no knowledge, no power, nothing inside me left. Nothing to do, and nobody to tell. And that was when I first began to think about suicide.

"I got there slowly, though. I think it was in steps. First, I realized how much better it would be to face a real fear, a concrete tormentor. Dealing with a known enemy, even a death with fixed timing, would seem like paradise by contrast. I came to dream of that death, in waking sleep, through the days and nights of fall, and into the depths of winter. And then I thought of taking violent control of that death, and being the one that could set the exact date and time in a final step that nobody could prevent, and so I would be taking back control over myself. And this, once I had conceived of it, was so desired.

"I don't know if I was depressed. I think that is just a word you use when you hear suicide. I know you like to use it in psychiatry, here in the West, in your West. And it's fine, you can call it depression if you like—of course I was not happy. But let me tell you another way to look at it.

"In the cotton fields of my home, our west, in what you call Xinjiang, the farmers have some problems with aphids, and in school the students who were interested in biology—like me—learned about the wasps that the government were bringing in to suppress the aphids. A lot of the Uighur kids interested in biology were guided into that field—the party was seeking modern kinds of employment for our people—not because they cared, really, but to prevent radicalization.

"Wasp warfare like this makes some sense—each wasp type is specific, restricted to its target species, so there is little risk of causing new problems. The female wasp injects her egg right into the aphid she catches; the egg goes through what would be her stinger, it's called an ovipositor—sometimes a paralytic is injected too—and then the egg hatches into a wasp larva that lives in the aphid, and grows, and partially eats away its insides, all the while careful to not damage the aphid's vital organs.

"Then the wasp larva breaks out of the belly of the aphid but still makes sure to keep it alive, and stays close underneath, and spins its cocoon with the aphid helplessly forming a living shield—the aphid is paralyzed but capable of a few simple movements if anything comes near, to protect its invader, to defend its killer—until the new adult wasp emerges from the cocoon, and the aphid only then is allowed to die, at last.

"So let me ask you: if that aphid could become aware, could come to understand its situation and choose death first, would it? Of course it would. And if the aphid could slowly emerge into consciousness, feeling the full agony of its situation as a human might, as it considered death, would you call the aphid depressed? I guess you might, but there would be no point—since no medication, nor any conceivable treatment, could be worthwhile, even if it could change the feeling inside.

"It doesn't matter; it's just words. I wanted death, and planned my death. That's what matters."

•

It was at this point that I began to realize the responsibility, and privilege, I had been given—to encounter this human being and her story. I did not merit this, to be the one hearing her, but fate had created an astounding crossing of major threads—historical, medical, emotional— at this moment in space and time, and so I could not cut her short when the appointment time had elapsed. I let her story unfold fully, her images taking form inside me, her experience linking to what I knew in science and medicine.

From the moment we first met, Aynur was completely at ease, and seemed impelled to share rich personal stories, adopting an interpersonal style more suitable for old classmates meeting at a school reunion. By itself this could be a red flag, regarding both the patient and the therapist—and their interaction—but I ended up finding no hint in her, or in myself, of what a psychiatrist could look for in such cases. For example, I never identified interpersonal patterns she had lived through that I might be evoking in her—a formative teacher, or an older brother, or a local doctor—nor did I perceive any pattern she might be bringing forth in me from my own life. This is always a risk, the patient and psychiatrist fitting into roles and calling up feelings from past experiences—often a problem, and sometimes a solution, in therapeutic relationships.

There were also no hints of personality or mood disorders in Aynur. Borderline and histrionic personality traits would be near the top of the list, in principle—along with hypomania, that stably elevated state on the mood-disorder spectrum—but there was nothing to corroborate these. Aynur simply told her intensely personal narratives with a natural frame of friendship, in as pure and engaged a social state as I had ever seen or imagined, with speech rich and anecdotes textured, somehow all in a language she did not know well, in a country she had known for less than a year.

Aynur seemed to me an archetype of the social state that our lineage might have evolved to allow, and as I listened I thought about the costs incurred—the metabolic bill paid each day, the brain resources allocated in each individual, for such a state to be possible—and also where this all began for the social mammals in our ancestry, perhaps in troops of early primates. The costs would have been substantial, I thought, since there is nothing as uncertain, and therefore as challenging to compute, as the social interaction in biology—not even the hunt of an unpredictable prey. The cat cannot predict which way the rat will turn, but there are not nearly so many possibilities as in a human interaction. And no hidden agendas—the rat wants to live, but what on earth does another person in a conversation want? And of course the rat can usually only express its drive to live in two dimensions, running on the plane of the ground. Likewise for the boxer, there is just a left and a right hand to worry about, and certain sequences and trajectories for each.

But the social brain needs a new mode of function, still requiring swiftness yet also operating along an enormous number of dimensions, running in a regime where a little bit of new information—any deviation from the current model, perhaps caught by and encoded in a few cells only—should be able to tip the observer into an improved model of the other individual, and the interaction into a better-predictive timeline. Yet the observer's brain also should not be overexcitable, and should in fact resist noise in the system that might cause switching to an incorrect model. It would still be important to suppress spontaneous ignition of a false perception, a harmful perspective that might arise from random sparks of neural fire.

As with anything in biology, the importance of such a process can be assessed by the consequences of its absence. We know the barrier, the lack of connection—distance and mistrust—when eye contact is even just a moment too brief. Yet a chilling effect also arises from eye contact held a fraction of a second too long, if not paired with a social signal of warmth. Temporal precision is clearly essential, as much for social interaction as for anything else in biology—severely pressuring the circuitry responsible for imposing the odd and ponderously slow pace of consciousness, that two-hundred-millisecond delay.

One possible solution, to speed this give-and-take, would be a premodeling, an unconscious gaming-out of events ahead of time in the brain. This feat could be achieved if the social being had many models of the world—and of the social partner—running at once, under the surface, that predicted the other's actions and feelings well into the future.

A key role—perhaps the major role—of the mammalian cortex might be to solve this prediction problem, to run models of the present and future, bringing together as much contextual information as possible from the world to inform these models. At the same time, the cortical system would have to detect with great sensitivity even small surprises—deviations from the current model that would indicate the need to hop to another. Running these countless models in parallel would make it unnecessary to calculate and spool up a whole new timeline in the conscious mind with each new bit of information, since each model would provide and prescribe actions, replies, forks into the future—moves and countermoves—over many timesteps, in a conditional hyperchess of the social mind.

The computational energy required in the brain, in order to constantly run these unconscious predictive models, would be vast. Perhaps this expendable element is the neural circuit-level resource exhausted quickly in the introvert, or in people—most people—who tire from prolonged social interaction. On the other hand, people with especially deep resources for this brain state would be the true extro-verts, thriving on constant human contact—like Aynur, as became clear even in the early going of the time we had scheduled, in what was supposed to be a quick and unremarkable evaluation, a check-in of sorts occasioned by her brief suicidality while she had been living in Europe. It was an interview unique in my experience—not just for the searing circumstances of her life, but for her intensely social disposition—and at the heart of it all was a human being who had wanted to die.

·

"There are two ways we seem to choose to do this," Aynur said. "In my hometown the buildings were not tall enough to make death from a jump certain, but in Kashgar they can do that, and certainly in Paris. The other way, well, *atlas* silk is so strong. I had many sashes, and one easily sets up bricks or books that can be kicked away under a rafter, or perhaps beneath a trellis in an outside garden.

"Why didn't I do it? My mother, I think. Even if I were forced to lose all my dreams as a scientist, and I could only have one piece of bread a day for the rest of my life, I would accept that, if I could be with my mom.

"The Parisians say they are more social than the Americans, and they are in some ways—they spend much more time with their family and friends. But nothing like the Uighur. You will laugh, but after my marriage I still slept between my two parents, in their bed, for months, as I had my whole life until then. This would be impossible in your West—unwifely, or worse. But that is how close we are. I could not end my life, because of family, because I could not harm those people close to me. I could not murder those relationships with my own hand.

"So I went on, alone in Paris, gnawed away from the inside, and then when I was somehow still alive after three months, in the abyss of win-ter, they released my husband, and he was able to contact me. Like all the young men there, my husband had been sent to a concentration camp. There might be another word for it in English, I don't know, since they were not killed, not really.

"When they released him he called me, we did a video chat. He was much thinner, with a shaved head, and his voice was very weak. I didn't know if he had been tortured, but he was so much more quiet, even more hollowed than I, and would not speak of what had happened. He told me he was being moved out of Xinjiang, to work in the cities closer to the coast. That was all he could say, he would be transported east, and he was not sure if or when we would meet again. So that's it now, he lives as a shell, making coarse movements.

"That's where things are still, mostly. This was last year, before I transferred my studies here from Paris, when the government was still denying these camps existed. This year, they are admitting existence, but they are calling them educational centers. People are sent there for failing to learn and speak the oath of allegiance in Mandarin. Or for being two-faced, as they call it—saying all the words correctly, but somehow not showing the right passion in their actions, or a deep commitment to the state.

"Oh, and they bulldozed the mosques in the town, while the young men were in the camps."

•

As Aynur was my last appointment of the morning I did not have to stop her to see another patient, only needing to sacrifice my lunch hour—an easy decision. My assessment had been long clear, and completed: she had problems in the past only—anxiety symptoms and an adjustment disorder due to extraordinarily stressful life events—with no current psychiatric diagnosis. In a patient with cognitive difficulty in domains beyond the social (Aynur had none, and was working toward a postgraduate degree in evolutionary biology), and who had shown certain facial features, I would have considered Williams syndrome, a chromosomal-deletion disorder. Despite anxiety and cognitive impairment, Williams patients can seem extremely socially adept—exhibiting expansive, rich storytelling and immediate personal bonding (though of uncertain depth) even with strangers.

Williams syndrome is still mysterious to this day, and still fascinating. But my clinical specialization was more focused toward care at the other extreme of social skill—treating human beings with brain states less inclined to, and less adept for, social interaction—on the autism spectrum. This was one of my two clinical passions (along with treating depression). From the earliest days of my practice, when I first emerged from residency training as an attending psychiatrist, the clinic intake staff were instructed to guide patients needing evaluation for a possible autism diagnosis to my care. I also asked the intake team to send me known autism patients who were challenging—people already diagnosed on that spectrum, but complicated for one reason or another, and referred by outside doctors (the same process that brought depression cases to my clinic). In this way, following the mystery of the underlying diseases, I found myself becoming a doctor who specialized in two disorders nearly untreatable with medicine: autism and treatment-resistant depression.

Knowing that no medical treatment existed for autism itself, I sought to help, in some way, a large and growing underserved population: adult autism patients no longer under the care of their pediatric physicians. These patients almost always suffer from treatable conditions occurring along with autism—comorbid conditions, as we call them, such as anxiety. My consideration in starting this clinic was that these disorders were often deeply affected by, and certainly framed by, the autism itself, and so could be best treated by a doctor with some specialization in altered social function.

Severe autism is defined by partial or complete inability to use language. But autism on the "high" end of the spectrum—socially, still the inverse of Aynur's state though with good language skills—also comes with its own challenges. Because these autism-spectrum individuals still have impaired social understanding, they can face a difficult conundrum in living their lives. With language ability and intelligence generally intact, and capability for employment typical (or even exceptional in the modern world), interaction with the broader community proceeds—but this interaction can be confusing and intensely anxiety-provoking, leading in some cases to serious new symptoms.

The social realm, and society in general—dominated as it is by the vagaries of human behavior—can be a mystery, even a minefield, for these conversant autism patients. How did that person know what to say just then? How on earth is consensus reached in a group? Where am I supposed to look as this person speaks? For these patients hell can indeed be—as Sartre noted—other people.

People are complex systems, but complex systems per se are not the problem for these patients, nor even complex systems that change with time—as long as the dynamics are predictable. Lines of code, a train moving along a one- or two-dimensional track on a timetable, the interlocking street architecture of a city—though complex, these can be appealing by virtue of predictability, especially to people living with autism. On the other hand, unpredictability—exemplified by social interaction—can be highly aversive, especially to those on this spectrum.

Understanding the precise sense in which social interaction can feel bad, I thought, will matter for the underlying neuroscience, and for helping human beings who live with autism (the entire spectrum all dwelling at the opposite pole of social aptitude from Williams cases, and from Aynur). Was social avoidance in autism not resulting from exhaustion of some computational or energetic resource, but rather arising from a fear of uncertainty or other people? Or perhaps there was something more subtle and hard to put into words at work here. For me, this latter possibility underscored the magnitude of the challenge of autism: how will patients already limited in linguistic expression tell us what is going on inside, if even we can't put it into words—and worse, if the words don't quite exist anyway?

I had long taken the opportunity to speak with my high-functioning autism patients who had good language skills. After building a therapeutic alliance over months of outpatient work and treating their comorbidities (as much as was possible), in follow-up clinic visits I would ask questions about the nature of their inner experience. But where to begin? I couldn't simply ask the patients to explain their autism. Instead I started simply and concretely—asking the patients about their experience of a single physical symptom. Of all the behavioral traits of the autism spectrum disorders, the avoidance of eye contact was to me the most arresting, and perhaps could be most illuminating: sometimes a brief flicker of contact, then the eyes flutter and flee like flushed quails—to the floor, to the side.

A patient named Charles gave me the clearest answer on this symptom. A young information technology specialist, he had what we used to call Asperger's syndrome—on the autism spectrum, but with excellent language skills—and extremely prominent avoidance of eye contact. In my clinic, I had spent two years treating his anxiety (successfully, in that he no longer suffered from panic attacks and workplace anxiety). But at the same time, there had not been one glimmer of change in autism symptoms, including his eye contact pattern. I asked him one morning, "What does it feel like when you do briefly make eye contact? Does it make you feel anxious or fearful?"

"No," he said. "I'm not afraid."

"Is it overwhelming?" I asked.

"Yes," Charles said, with no hesitation.

"Tell me about that, Charles, if you can."

"Well, when I'm looking at you and talking, if your face changes then I have to think about what that means, and how I should react to that, and change what I'm saying."

"And what then?" I gently pressed. "What exactly makes you look away?"

"Well," Charles said, "and then that overloads. It overloads the rest of me."

"So it's like too much information, and that feels bad?"

"Yes," he said right away, "and if I'm looking away, it's easier."

For me, as a neuroscientist and psychiatrist, this was a transcendent moment. Though sitting before me was a patient with severe eye contact avoidance who was clearly susceptible to anxiety, I had been allowed to hear something that few scientists were privileged to know unequivocally: that the eye contact issue was not due to anxiety. This conclusion was strongly supported by the completely separate fates of the two symptoms (anxiety and eye contact) under my treatment—one was cured and the other was utterly unaffected. For this patient, at least, this separation between anxiety and eye contact was also directly confirmed by his own description, in words of a human being positioned on the autism spectrum so precisely as to be strongly symptomatic, and yet also fortuitously verbally expressive enough to share his internal experience. In some ways this single moment justified for me the entire career progression I had taken, all the extra years of both MD and PhD training, all the pain and personal challenges of internship, all the call nights as a single father, worrying about my lonely son. This alone was enough.

Instead of anxiety or fear, a much more interesting, and subtle, process seemed to be at work. Charles's brain was detecting its own inability to keep up with the social data stream—while aware that it should be keeping up, that this was a situation where processing the data was essential. And more: his brain had created a link from that informational challenge to a negative-valence subjective internal state, a state of feeling bad.

Mysteries remained, as always. For example, was that negative feeling innate or learned? His link from high information rates to feeling bad could have been taught by life, conditioned over time by repeated and emotionally difficult failures of social interaction. Or was the aversion present from the beginning of his life, without training? Was feeling bad an evolutionary mechanism to help people dodge the torrent of data, guiding them to disengage from full participation in situations where correct responses to the data would be expected by others, and failure thus socially consequential—harmful, even? Was the unpleasant feeling then just triggered by the unpredictability of the data streaming in—essentially a response to the high bit rate of information itself?

This was an idea that might matter, a possible insight that had been gifted from just the right patient—born on the pole opposite to Aynur, but still verbal enough to tell us his story.

•

"It is so unfair," Aynur continued. "We are a nice people. We are not only close to our families. Guests in our home, we put at the head of the dinner table. Any visitor, no matter who, receives this position of honor. That could never happen here in California—and not in France either. It's funny for me to see how you are. It is like you are afraid the guest will take the home away from you.

"Are you really worried about that? It is your house. Nobody is going to take it from you. If we have a guest, we give them the best seat for one evening. And it makes a strong bond. So much strength lies in the gesture, costing nothing and creating a deep connection that will last forever.

"I wonder if this part of our culture is interpreted as weak. But it's not just the Uighur, all the communities do this, the string of settlements all across the middle of the continent—we call it the Silk Road, and you call it that too, I think—but I think this is how we survived, because we could be a social culture. And we are strong in many other ways—not just in social bonds. When I was thirteen, I fought seven Han girls by myself.

"We were in our dormitory and they were talking; I knew they didn't think I could understand them. I was always so much better at understanding languages than people imagined—I learned Mandarin, and French, and English, in only weeks each it seemed, all just by hearing and watching. And these girls were complaining that someone had left a food dish out in the common area, and they were blaming me. And then one who was standing in the bathroom in front of the mirror, brushing her hair, said a terrible thing about my family, my loved ones she had never met, how my mother smelled. I leapt down from the bunk and dragged the girl by her hair out of the bathroom. The others all jumped on me but were surprised, even I was surprised, that I was stronger than all of them together. I had no idea until that moment that my legs were so powerful. They fell against me like raindrops, a storm that passed quickly—and I never heard a rude word again that year.

"I feel guilty today about those girls. I was the one who became violent first. It seemed I had to defend my family, but now I am twice that age, and I see they were just kids. And maybe I made things worse, maybe I harmed their perception of my culture. The Han are good people too, and I don't blame them for their government. But I wonder now if there even is a path forward for them, for their country, to move in a new direction, to no longer be part of such a system. Can they evolve away, or have they fallen into something from which there is no escape?

"I studied more biology of the aphids after I started my master's degree, and I learned the history of the wasps, incredibly successful animals which come in more species than do any other order of animals on earth. Where does this success come from? Did you know that ants, bees, wasps, and hornets all started from the same wasp ancestor, in the time of the dinosaurs, when one little plant-eating fly, like a sawfly, was born with a strange mutation that made its egg more layable into animals, right through its ovipositor, that stinger-like egg-laying tube? And from that one moment an incredible radiation of animals from one ancestor occurred, because it was so powerful to be able to lay eggs in any living thing—into a spider, an aphid, another wasp.

"That incredibly thin, hairlike wasp waist—the tiny connector linking one part of the body to another—was created by chance mutation. Then natural selection took over, accelerating the expansion of wasp species—the spread of waspiness—using the waist to allow body contortions to position and guide increasingly long ovipositors, to get to beetle larvae deep in trees, to body cavities deep within large caterpillars.

"But the most surprising final part of the story, that matters here, is that several branches of the wasp family—ants and hornets and bees, all the social groups—later reverted away from this life cycle, completely abandoning the parasitoid egg laying into other organisms that had made them who they were. Complex body parts are easily lost in evolution if not needed, and are never regained; it is rare for organisms, once parasitic, once evolved to extremely reduced body plans, to escape that evolutionary pit. But these did escape in a different way, through being social, by dependence on each other. They had found a way of living together—and commitment to this social mode had set them free.

"They still retain that severely reduced wasp waist—you can see it in ants, and so clearly in the yellow jackets you have around here—though they don't need it to be so thin anymore. The wasp waist is a mark of their ancestry, but their ovipositors have converted to stingers to defend their family, and they use powerful social structures and bonds to take care of their larvae, and no longer need to place their young within another living being.

"Do you know, it took fifty million years for wasps to figure out how to live in groups, even family groups? Social behavior is hard. Before that, just seventeen million years had been spent inventing the wasp waist, and then only another thirty million had been needed to turn the ovipositor into a stinger (by the way, this is why most bees are female: the stinger arose from that female reproductive organ, the ovipositor, and so only females can defend the family)—but even then, the social challenge had not been solved yet.

"After evolving the behavior of paralyzing hosts with venom, and laying eggs in or near them—wherever the host animal happened to fall—fifty million years were then occupied by developing increased levels of transporting the paralyzed host to a safe concealed location, and building a nest for the young to develop, and expanding to other kinds of food resources needing more work like pollen and leaves, and finally defending the nest as a group, as a family.

"Social behavior is rare, and a lot of things have to come together to make it work—beginning with prolonged care for the young, but then success is still contingent on many other factors, all somehow needing to be satisfied together—like having a sting ready to defend the big investment of the social group. And when everything is in place and working, a whole world—*the* whole world—opens up."

Here Aynur paused. This was rare for her. I uncrossed my legs and sat up a little straighter, folding my hands together in my lap.

"I had a difficult dream about a baby," she said finally, "after I came to California." She seemed to be struggling with the memory.

I gave her some time, not wanting to risk redirecting her flow even slightly. As I waited—not being an insect expert—I wondered about mammals, whose close parent-offspring interactions could have served as a guide for the creation of social behavior in our lineage as well. That same year of 2018, researchers studying parenting in mice had used optogenetics to deconstruct this complex interaction—finding individual neuronal connections governing its sub-features, including projections across the brain that provided motivation to desperately seek out and find the young, versus other projections that guided actual acts of caring for the young—with each action powerfully governed by a different connection across the brain, radiating out from an anchor point, just as we had discovered for assembly of the diverse features of anxiety five years earlier.

The intense and ancient parenting dyad had created these neural circuit foundations, which could have been used again for new kinds of social interactions. An insect that can care for its young perhaps more easily becomes an insect that can care for its nestmates—and the same idea would apply to a mouse, or an early primate—by repurposing the same neural circuits. And all those techniques of caring for young, of good parenting itself, could have arisen first by circuit repurposing as well (such recycling seems to explain so much of evolution)—here by inserting the other individual, the offspring, onto one's own need and motivation structure, creating a simulation dwelling within, as a trick to use the self's internal processes to rapidly model and infer the other's needs.

Yet nonfamily social behavior would seem to be fundamentally more complex, since in families the motivations of both caretaker and offspring remain—for the most part—certain and constant. In contrast, the most interesting challenge of true nonfamilial social interaction is that of keeping up a rapidly shifting internal model, changing every few hundred milliseconds, to predict actions of another being with highly uncertain goals. And though many mammals do show nonfamilial social behavior, it is a fragile construct; from lions to meerkats to mice, social mammals are often only moments away from murdering each other.

"In the dream I was myself," Aynur continued finally, "a regular human being like you. I was also a parent, which was strange—I've only carried a teratoma in real life. But babies were also different in the dream—born smaller than a thumb, more like a pine nut, tiny and nearly hairless, like the marsupial babies that first emerge almost like small drops of pink liquid with forepaws, born with just enough dexterity to pull themselves along their mother's belly fur, to find milk and survive.

"In the dream, all human babies were like that, except even more helpless. If you were a human parent in this world, of course you had no pouch, and no fur on your belly, and the deal seemed to be that if you had a baby, you just had to carry it in your hands.

"The babies were so small they all looked similar, like embryos do. But if you had some, you knew, you definitely knew yours—partly because you could never put your babies down, so you always had them with you, and carried them on your journeys, wherever you weaved your way, along the lakeshore, or through the taiga, carrying your little droplets of human warmth.

"In my dream I lost my baby on the forest floor. I don't know how it slipped away, or when. I tried to search along the path I had taken, but the ground was covered in the dying foliage of late fall. I frantically sifted through the mat of fallen needles and bark, but it was hopeless: there was so much space to search, and my baby was so small.

"My child was helpless and cold and dying somewhere on the ground, apart from me. As I searched I could feel a fine thread connecting us—the baby was me, a part of me, apart and needing me, though I could not see where in the outside world the thread projected. But within me, the loss had a definite place, a position in space that I could feel. It was in my chest, behind my breasts, in those deep muscles that sweep the arms. The inner feeling, of the loss of a child, had been mapped there somehow—that was where evolution had set this feeling, this was how it should feel, to make me do what needed to be done. It ravaged me as I dug, and drove my arms to seek the piece of my heart that I had held so long. It was a gap, it was a savage gape, and it made me dig."

•

Aynur's comfort with complexity was not only in the social domain. She seemed to synthesize every stream of information available, of all kinds—her dreams, her memories, her science. Everything was related, and it all mattered together, and she wove it all together effortlessly. On the other end, perhaps in a related pattern, the social information stream that Charles found overwhelming was not the only kind of information that was aversive to him. As with many people on this spectrum, he had trouble with unpredictable events in the environment more broadly (sudden sounds or touches, for example, he found more distracting than did most other people—these were even painful). And so the positioning of different people on this autism spectrum could boil down to processing all types of information—not only social; symptoms were perhaps just clearest in the social domain due to its uniquely high rate of information flow.

Thinking of the rate of all information as the challenge in this way, rather than just social information, could also usefully explain unpredictability as a key problem in autism. Only unpredictable data is really information; if a person understands a system to the point of predicting everything about it accurately, then it is impossible to inform the person further about the system. The challenge with autism, then, seemed to be about the rate of information itself.

I didn't know when treating either Charles or Aynur, nor do we know now, exactly how information is represented in the brain—at least not with the same code-cracked certainty with which we know how genetic information (at the most basic level) is encoded in DNA. But we do know that neuronal information is transmitted by electrical signals moving within stimulated cells, and by chemical signals moving between these cells. And many of the genes linked to autism are related to these processes of electrical and chemical excitability—encoding proteins that create, and send, and guide, and receive the electrical or chemical signals.

So the genetic evidence I knew was at least consistent with the concept of altered information processing in autism. That idea alone is not specific enough to be useful in guiding diagnosis or treatment, but numerous other signs and markers point to altered information flow in autism. Averaged across the population, people on the autism spectrum exhibit signs of increased excitability, or triggerable electrical activity, of the brain—such as epilepsy, a form of uncontrolled excitation taking the form of a seizure. And in measuring brain waves with the electroencephalogram, or EEG (external electrodes that can record synchronized activity of many neurons in the human cortex), certain high-frequency brain rhythms called gamma waves, which are oscillations coming at thirty to eighty times per second, show increased strength in human beings with autism spectrum symptoms.

As a result of this evidence, it had been widely speculated that a unifying theme in autism could be an increased power of neuronal excitation—relative to countervailing influences like inhibition. This hypothesis was well articulated and appealing to many in the field, in part for its flexibility, since diverse mechanisms—from alterations in neurochemicals, synapses, cells, circuits, or even whole brain structure—could implement such a change in the balance of excitation and inhibition. For example, since the brain contains both excitatory cells that stimulate other neurons and cause increased activity, and inhibitory cells that shut down other neurons, one attractive form of this hypothesis would be that autism symptoms could arise from an imbalance between excitatory and inhibitory cells themselves, specifically favoring the excitatory cells.

But how could such a hypothesis of excitation/inhibition balance be tested? Despite availability of clinical strategies to dampen overall brain activity, such as medications to treat seizures and anxiety, these drugs (for example, a class called benzodiazepines) turn down the activity of all neurons—not just excitatory cells.

Just as the excitation/inhibition balance hypothesis would then predict, benzodiazepines are therefore generally not effective for the core symptoms of autism. Autism is clearly not just increased activity in the brain. Charles, for example, who suffered from anxiety, had received a benzodiazepine prescription from me for many years, but this treatment—as I expected—had not touched his autism symptoms at all, despite eliminating his anxiety.

The cellular formulation of the excitation/inhibition balance hypothesis, untestable for so long, finally became accessible with the advent of optogenetics. If the relevant imbalance in autism—at least in some forms—involved excitatory and inhibitory cell types, optogenetics could be ideally suited to test this idea. We could increase or decrease excitability of the excitatory or the inhibitory cell types—in targeted brain regions such as the prefrontal cortex that handle advanced cognitions—using the microbial light-activated ion channel genes called channelrhodopsins, and with fiberoptics to deliver laser light.

Mice, like people, mostly prefer being with each other, even in pairings unrelated to kinship or mating; rather than being alone, they will generally choose an environment that contains another (nonthreatening) mouse. Mice also seem to express actual interest in each other, with prolonged bouts of social contact and exploration. And the human mutations known to cause autism, when mimicked in mice using genetic technology, can cause disruptions in this mouse sociability.

So with the generalized success of optogenetics in mice by 2009, it became apparent that this technology might be used to help illuminate mammalian social behavior. In 2011 we indeed found that optogenetically elevating the activity of *excitatory* cells in the prefrontal cortex caused an enormous deficit in social behavior in adult mice. Importantly, this intervention did not affect certain nonsocial behaviors such as the exploration of unmoving (and therefore quite predictable) objects.

The effect was specific, then, and in the right direction as predicted by (and therefore supporting) the cell-type balance hypothesis. Even more intriguingly, and fitting the hypothesis as well, optogenetically elevating the activity of *inhibitory* cells in the same already modified mice, to restore cellular balance, corrected the social deficit.

Crucial to this experiment had been our creation of the first red-light-driven channelrhodopsins to complement the known blue-light-driven versions. This advance allowed us, in 2011, to modify the activity of one cell population (excitatory) with blue light, and another population (inhibitory) with red light, in the same animals. That experiment had shown that elevating excitatory cell activity could cause social deficits in healthy adult mammals, and that this effect could be ameliorated by elevating inhibitory cell activity at the same time, to rebalance the system.

In 2017 (in a narrow slice of time after I had treated Charles, but before I met Aynur), we brought our approach to mice that were not typical—that had been induced to carry mutations in genes matching those found in human autism families. These mice (altered in a single gene called *Cntnap2*) had inborn deficits in social behavior compared with nonmutant mice. We found that this autism-related social deficit could be *corrected* by optogenetic interventions that were logically the opposite of what we had done to *cause* social deficits in 2011. Either increasing activity of inhibitory cells, or reducing activity of excitatory cells, in the prefrontal cortex (both interventions would be predicted to reduce cellular balance back toward the natural levels) corrected the autism-related social behavior deficit.

Beyond this proof of principle—causally testing the cellular balance hypothesis—we were intrigued by the demonstration that for these so-cial deficits, both cause and correction could be applied in adulthood. This was by no means obvious, and certainly the result could have turned out otherwise. It might have been that the relevant imbalance occurred only at some inaccessible and irreversible step earlier in life. If this were the case, the insight would still be valuable, but treatment interventions would become much harder to envisage. Our findings did not rule out any possible contribution from before birth, but did show that at least in some cases, action in adulthood could be sufficient for both causation and correction of a social deficit.

These results—turning social behavior up or down by shifting the balance of excitatory and inhibitory cells—also had illustrated the broader worth of a particular kind of scientific process, beyond the in-trinsic value of the scientific finding. Here psychiatry had helped guide fundamental neuroscience experiments, which in turn had helped il-luminate processes that could be going on within unusual human minds in the psychiatry clinic—coming full circle in casting light upon clinical moments as emotionally complex and intellectually profound as Aynur's immersive storytelling.

.

"I know we've gone an hour over our time," Aynur said, filling a pause that I just realized had existed, in the moment of its ending. "I'm sorry if you missed your lunchtime. Thank you for listening—I just wanted to explain. The French doctors wanted me to follow up here, but I'm not suicidal now. I had a time of weakness, that's all.

"I don't mean to be too dramatic about any of this, but just to say: I could become weak like that again. Now I know I need my family, and I cannot live without them. These bonds that created the human way of life, that maybe allowed us to survive, also might have left a vulnerability. I don't mean to say all of us would react the same way, but I know I never felt the weakness so strongly as in those three months, I was so nearly destroyed by something not related to food, or shelter, or even reproduction. I almost died, even though I could easily have found ways to stay in the West, with new friends, and new partners.

"I still could. There were men that looked at me. There was one man that I looked at too.

"We met and talked one night in a café by the stadium. Things seemed about to explode. How to describe it to you? Eruptive, I want to say, but I don't know if that's a word. Brimful? So many possibilities. I wasn't thinking in English then—that was more than six months ago—but it doesn't matter, none of the languages I know have the right words.

"Nothing happened, anyway. We just drank coffee from clunky purple cups. And I realized as I walked away, after, that our social bonds only reinforce a strength we already have built in.

"I knew something the man I had coffee with didn't know, that social structure came only after the venomous sting. Evolutionary biologists think that having such a sting was crucial for allowing evolution of social behavior in the wasp lineage, by providing a remarkable level of defense for what had been such a small and fragile animal. And I agree—you can only be social with strong weapons to defend your nest and young. That strength can free you from harming others. The need to connect with others *is* strength—not weakness."

Extroverts like Aynur, and natural politicians with near inexhaustible sociability, draw energy from conversations and avoid being alone—an inversion of the value system of those on the autism spectrum. And like Aynur and Charles, many people favoring one extreme of social intensity can find a coerced exposure to the other extreme unpleasant, like nocturnal mammals forced out into the midday sun.

Evolution has helped the nocturnal mammal to find daylight aversive, because this negative feeling makes the right behavior happen—which is to retreat from the light and await conditions that are more suited to the animal's design and hence less dangerous, and more rewarding. It is similarly possible that social or nonsocial brain states could be harmful if present under the wrong environmental conditions, which might help the mismatched condition (over the long timescales of evolution) become associated with aversive or negative feelings.

Just as different survival strategies are suited for nocturnal versus diurnal life, there may also be fundamentally distinct brain modes for different rates of information processing—each mode of value, but not mutually compatible (at least, not at the same time). The mode of dealing with a dynamic, unpredictable system (exemplified by social interaction) may be incompatible, or at least dwell in tension, with another mode that we need at other times. This second state would be one that allows us to quietly evaluate an unchanging system—a simple tool, a page of code, an algorithm, a calendar, a timetable, a proof—anything static and predictable, for which the best strategy toward understanding is to take the time to look at the system from different angles, with confidence that it will remain unchanged between one inspection and another. Brain states differentially suited for these two distinct situations might need to be switched on or off from moment to moment (with relative state preference tuned across millennia of evolution, and with variation from individual to individual in the strength and stability of each state).

Our optogenetic excitation/inhibition results were later replicated in independent mouse lines, but a key question remained: was there a link between this cellular imbalance shown to be causal for social deficits in mice and the informational crisis as experienced by Charles (and perhaps others on the autism spectrum)? Optogenetics helped unearth an idea for how this linkage might work; in our initial excitation/inhibition paper in 2011, we also had reported that causing high excitability of prefrontal excitatory cells (an intervention that elicited social deficits) actually did reduce the information-carrying capacity of the cells themselves, in a way that we could measure precisely, in bits per second. Thus, the very same kind of altered excitation/inhibition balance that disrupted social interaction was also making it harder for brain cells to transmit data at high information rates—corroborating what Charles had described for us in his account that the information coming in through eye contact *overloads the rest*.

Another remaining question was the origin of the aversive quality of information overload—so powerfully experienced as unpleasant by Charles and others on the spectrum. Being unable to keep up with social information feels bad in these individuals, but it is not obvious why. The information overload need not have any emotional valence at all, or perhaps even could have been positive—a sense of freedom upon realizing that one cannot keep up, with solace and a kind of peace in the resulting isolation. Here, though, I did understand, in part through listening to my patients, how difficult it could be to make your way through life with others constantly expecting higher social insight than you could routinely provide. And so the aversion might have been socially conditioned, learned through a lifetime of mildly stressful interactions, devastating misunderstandings, and everything in between.

But instead of this aversion needing to be learned, could information excess be innately aversive in itself, when above a person's carrying capacity? Certainly everyone, from the typically social to the simply introverted to the autism-spectrum individuals, can experience aversion after prolonged social interaction, when social circuitry becomes exhausted to some extent. It might make sense evolutionarily, in a long-social species like our own, to have developed a built-in aversion mechanism, providing motivation to withdraw from important social interactions when the system is fatigued, and likely to begin causing errors of understanding or trust.

•

"One more thing," said Aynur, as we stood up together—I thought I had initiated the action, finally needing to prepare for my one o'clock, but she had responded so quickly and closely that we were moving in perfect synchrony and then I wasn't even sure who had started it. "I know they just wanted me to have a onetime evaluation with you, and so we probably won't see each other again—but you had asked earlier, when we were talking about my family, how we made the silk to get those colors, and I didn't get back to that.

"This part is really interesting. I remember when I was little, I most loved the light-pink form of the tamarisk silks. It made me feel like seeing the flowering tree in person; the color seems so delicate but the silk is strong, all just like the tree. I don't know if you've ever seen one. The tamarisk is such a wonderful bit of life. A desert fir, evergreen but also colorful.

"By the way, there are wasps that lay eggs in the tamarisk tree. Then a new kind of wood forms around the egg—a growth, a gall: a twisting ball of nut and root. It's like a teratoma, but it doesn't hurt the tamarisk. There is no need for the tree to fight it.

"I was reading the other day, the tamarisk is now an invasive species here. You call it a salt cedar—I like that name. People say it was first brought over here from Asia just for decoration, and now it's taking over parts of the American West. The tree thrives in salt, and makes the soil salty too, which gives it an advantage over the willows and cottonwoods.

"In some areas around here, hikers are apparently being asked to pull up the salt cedar shoots wherever they are seen, to protect the native plants and animals. Birds are losing the trees they used to live in—but it seems doves are okay with nesting in the tamarisk—hummingbirds too. In other places people have given up fighting, and now just let it go. So there's a flood of salt cedar color in parts of the western desert. I saw a picture—you really need to see it. I wish I could show you.

"Anyway, what we do for silk—I can only tell you about the traditional way my mother taught me, how we do it slowly, by hand. I don't know how it's done when mass-produced. We sort the cocoons first, and for the stained and odd-shaped ones we have to boil them; they all change to the same shade in water.

"Then we stir with a stick, separate the threads, and twist the threads into strands; we need a few dozen threads to make a strong strand. To color, we dip each bundled strand into different pigments, dyeing one strand at a time. I remember this all as very slow, especially for those fine pinks and purples, the pale and light colors of the tamarisk tree."

.

An ever-growing share of all human interaction appears to lack the full vivid color of natural social information. By suppressing rich social multidimensionality, we relieve ourselves of its mental burden (though we may come to miss, or even crave, this burden once discarded). We suppress the visual stream of information on the phone, or simplify the entire social data stream using emails and posts and texts; each of these methods for reducing data-per-interaction confers a kind of insulation and enables a higher rate of individual social events, if desired (though allowing more frequent misunderstandings).

The trend toward increased social partners and contacts, with fewer bits of data transmitted at each contact, may have already reached a practical limit, approaching a mode of one bit per interaction (liked or not). That remaining bit can still be imbued with immense intensity, arrogating attention, driving passion and intrigue—because the bit is charged with social context and our imagination: that is, with premade models in our cortex, ready to run. Human connection in some form is now possible through only a few words or characters, even a binary flip of a switch—obviating some of the pressures of social complexity and unpredictability.

We could perhaps now relax categorizations of sociability (as a little outdated) that define what is healthy or optimal based on only the typical high-information-rate in-person social interaction. People with autism (at least on the high end of the spectrum) can seem more socially adept if the interaction is moved out of real time—to low bit rates, as with text. Though any interaction style still risks errors and misunderstandings, communication can seem improved if given the grace of time.

The bits to be transferred can be prepared at leisure, and then discharged when ready with a tap; no reply is needed immediately. These bits may then be placed in the broader context by the recipient—over minutes, hours, or days—to be evaluated from different angles. Possible replies may be considered, and scenarios run forward like a chess match for two or twenty moves, off the clock—until a reply may come, a tap or two when ready: Morse for the late-modern human.

The autism spectrum, then, need not be seen entirely as a "theory of mind" challenge—which has been a popular and helpful idea, holding that in autism there is a fundamental problem with even conceptualizing the minds of others. Instead, the bit-rate-limitation idea (which optogenetics has helped reveal) would perhaps fit more fully with the experiences of many patients, who are capable enough but require time to run their models, to fit their own carrying capacities.

Psychiatry and medicine broadly—though still constructed around interpersonal communication—can survive and operate well with much less social information than the traditional face-to-face interview provides. I came to this understanding first as a resident at the local Veterans Administration hospital, where (under the relentless pressure of overnight call shifts) I found that the unique human connection needed in psychiatry can form first through a thin audio channel, the low-dimensional projection that is a phone call, if extended in time.

I then rediscovered this for myself, also in a time of necessity, as an attending during the global coronavirus pandemic of 2020. Emergency psychiatry, I saw again and again, though it somehow surprised me each time, can be carried out with precision even over the phone, through that lonely single line.

The Veterans Administration hospital rises like a mirage out of grassy foothills near the university. An oasis of contradictions, this VA system inspired Ken Kesey's *One Flew Over the Cuckoo's Nest*, but is now largely staffed with university-affiliated academic physicians at the leading edge of the field—and so to this day, the VA still evokes simultaneously both psychiatry's prescientific troubled distant past and the neuroscience-driven promise of its near future.

The on-call psychiatrist at the VA is dubbed the NPOD (neuropsychiatrist on duty). The main duties of the NPOD (one resident for the whole hospital, all night) are wrangling emergency room admissions, responding to consults from inpatient services, and caring for the psychiatry inpatients on the locked units. A major side job, however, is fielding calls coming in from the outpatient community, throughout the immense catchment area of this flagship hospital encompassing all the military veterans who might be phoning in from home—especially those suffering from PTSD (a common and deadly disease that is often resistant to treatment by medication).

Page the NPOD: an invocation when all else has failed. In the midst of other emergencies, the NPOD receives a call forwarded from a veteran beyond the walls of the hospital: a reaching-in from a human being who is jangled and guilty and helpless, needing only to talk with someone, anyone, who might understand. I found these calls could require an hour or more to work through. Less time would be taken in person, but a different mode was needed for these purely auditory conversations that were still sensitive and vital, with the gray specter of suicide looming.

When the call would come, seemingly always near three o'clock in the morning—perhaps in the middle of a chaotic scramble from the inpatient ward to the emergency room, or sometimes just as I was going to try for a few minutes of sleep in the barren housestaff call room—early in my training it was hard to suppress feelings of anger, especially since there was no concrete goal for the call, at least that a combat veteran could typically describe. The patient just needed to talk—and so I learned to transform myself from efficient physician to purely empathic partner. Both veteran-as-patient and myself-as-NPOD, I came to realize, were fighting a new battle in different ways, each trying not to bring feelings from prior personal trauma into the present, to not transfer presumptions and imputations from one context into another.

I would often field these pages in the call room, curled up for hours on the impossibly hard and narrow plastic mattress—still in scrubs and ready for any urgent summons to the locked unit for patients with chest pain or needing restraints—but under a thin hospital blanket to ward off that bone-chilling pre-morning despair, phone braced uncomfortably between cheek and shoulder. Not a propitious arrangement for deep connection, yet somehow by the end of each call, the patient and I could usually both move on—to the next interaction, the next challenge, or perhaps even to a shallow snippet of sleep—with a kind of peace, a gift of warmth from another human being, after a true social interaction drawn out across the line.

The coronavirus sweeping over the planet, years after my VA service, coerced a retelling of this story in a new way. As populations from city center to countryside became fragmented, by design, to manage the contagion, many human interactions either were forced to play out at a distance or were simply sacrificed. The culture of traditional psychiatry thus seemed initially vulnerable. Video and phone appointments (desperately needed as a replacement for clinic visits during the crisis) were for the first time widely approved and scheduled; this normalization of virtual psychiatry interactions had long been possible but resisted by the established clinical structures for an undeniable flaw: lacking the full information rate of in-person communication.

Younger patients were immediately at ease with video appointments via the Internet, considering this interaction to be as natural as any other (and even preferable), but some of my older patients were uncomfortable with the idea and preferred the telephone. During one of these audio-only visits—with Mr. Stevens, a man in his mideighties I had known for years—I was startled by the immediate reactivation within me of that intensity of focus and feeling, all centered on the spoken word: that purely auditory information stream, that thin squiggle of time-varying sound, which by necessity had guided so much of my psychiatric caregiving during call shifts in residency.

Mr. Stevens had relapsed into depression four weeks prior (before the COVID-19 pandemic took hold in California), at which time I had bumped up the dose of one of his medications. Now as I exchanged pleasantries at the start of the phone call (taking time even before discussing his disease symptoms, while knowing that if suicide were a risk, I would never see him face-to-face in time), I noticed that I had adopted that familiar life-or-death focus on his timbre and tone and pauses and rhythms that I had learned with the veterans at the VA—and that I already knew all I needed to know about his mental status. By the time we got around to his actual description of symptoms and feelings, I found we were only confirming what had already become clear to me, quantitatively and with certainty: that his depression had lifted by about 20 percent.

The most socially adept among us do this all the time—those beyond my own capability, who without effort or training or delay can see through the vast avalanche of social data at just the right angle to find unerringly the meaning of the moment. But every part of us contains our whole, if reflected upon. Even with little carrying capacity, connection still comes, with time.

•

"I feel like I want to tell you more," Aynur said as we stood in the doorway of my office. The hallway was quiet, and the carpet looked drab and dim. "It would be nice to talk again, but I guess I know we won't ever. I am sorry. I know there is no time, but one more thing: I had a final moment you should know about, the morning I was to leave Europe. Not looking at a man, but looking at a girl.

"It was six in the morning, and I was gazing out from my small loft window, drinking the last of my tea, preparing to leave for the airport, and taking a minute to pause and reflect—to pay my respects, in a manner of speaking. There was not really a view of the city, just the gray apartments across the alley, but I still felt this was a goodbye to Paris, a quiet moment of homage. I had learned and changed a great deal, and the French doctors might have saved my life. As I looked out into the light mist of morning, toward the tenement across the way, a ten- or eleven-year-old girl in a hijab emerged alone onto the narrow balcony.

"I had seen her and her family before, in passing, in occasional glimpses, the snapshots one gets. She seemed to have a little sister, and they lived with their mother and father, who wore traditional dress, not typical French style, though I did not know the country. But this was much earlier in the day than I had ever seen her before, and she was alone. She looked out toward the east, followed by a quick glance back to the darkened apartment. Her face was set and serious; she was not there to enjoy the sunrise.

"Then she moved to the balcony, to the edge, and turned her back to the sun, facing west. I held my breath—for her. I had envisaged myself jumping, like this, so many times—while looking out this very same window.

"She took out a phone, and hunched over for a moment, then straightened up and held it out in front of her. In a moment her whole demeanor had changed; she had become a movie star, her face shining with fierce glamour. It was just a selfie.

"She then returned to her position hunched over the phone, looking at the image. She remained this way for almost a minute, and then quickly looked at the sliding glass door back to the house, which she had left partly open; all seemed well for her, remaining dark within.

"For the next ten minutes I watched, enthralled, as she flipped back and forth between the two positions. Her next selfie was another joyful one, then one with a silly duck face, followed by one with her tongue stuck way out, peace-sign fingers forming a horizontal V wagging just below her chin. After each one, she bent abruptly into an intense state of frozen scrutiny. Her focus, her intensity, was impressive. This seemed a rare stolen opportunity—perhaps her mother was in the shower and would emerge any moment. Back and forth she went, on and on, almost puppetlike in the stereotypy of her transitions. I had always seen her, interpreted her, as a small child with her doll, but here she was being jerked back and forth by something else, a new drive—whipped by an unchildlike need.

"Eventually she was satisfied. She slipped back inside, and was gone.

"I felt profound sadness, and joy, and jealousy, all together. Is there a word in English for that? I've felt that before, those three things together. There should be a word. All three basic layers of emotion, down and up and sideways, all wrapped up in a tight and disorganized little ball.

"The jealousy—though we shared faith, gender, youth . . . our cultures were still so different. She was still blessed, gifted, able to begin a journey I could never take. I was bound too firmly to my own, to my trapped and now tortured people.

"My joy came from knowing this was the outset of her journey, that she was setting off from her family's homeland, preparing to weave a new fabric of her culture, traveling down her own road to autonomy.

"Though moments like this, of course, must occur thousands of times a day, every day, around the world, my sadness might have come from realizing her parents would never know what had happened on the balcony, in the way I did, as a complete stranger; this was a poignant hidden moment of a girl separating from her mother's hand that would never be shared. The sadness also, I guess, came from my own selfishness—from feeling connected to this girl in many ways, but realizing I would never get to know her deeply. I was still feeling vulnerable, or empty—from my teratoma. From everything.

"She was found and lost to me almost at the same moment. I never existed for her, and never would, and she ended up only a sort of cross-thread in my life—marking a moment—though with a thread that is strong, and durable, like in that rough ribbon you have here with ridges and gaps alternating, called grosgrain, where the weft is even thicker than the warp.

"It is strange to say, but the thickness of her one thread formed a gap around which nothing else can come close. I came to know her deeply, though it took only a few minutes, and now she feels lost to me. I don't know how, but I might need to find my way back to her."

CHAPTER 4

BROKEN SKIN

> As willing to feel pain as to give pain, to feel pleasure as to give
> pleasure, hers was an experimental life—ever since her mother's
> remarks sent her flying up those stairs, ever since her one major
> feeling of responsibility had been exorcised on the bank of a river with
> a closed place in the middle. The first experience taught her there
> was no other that you could count on; the second that there was no
> self to count on either. She had no center, no speck around which to
> grow.
>
> —Toni Morrison, *Sula*

Henry, nineteen years old, had been found rolling naked on the aisle of a county bus. When the paramedics arrived, he told them that he was imagining eating people, and saw visions of himself consuming flesh and bathing in blood. But after his swift transport to our emergency department by police, Henry gave me, the on-call psychiatrist summoned to evaluate him, a more relatable story instead, with more universal themes. He described a lost love that had brought him to despair, to the aisle of the bus, to suicidal thoughts, and to me.

Not even guessing at his diagnosis yet—there were still too many possibilities—I let my mind work freely, picturing the scene as Henry described his magical first moment of romantic connection from three months ago. In her short fur-lined coat, Shelley had knelt on the torn vinyl seat of the church field-trip bus, leaned close, and kissed him—just as a sunbeam unexpectedly broke through the canopy of trees and fog. More used to the pervasive chill of early spring among the coastal redwood groves, he was surprised and enthralled by the sudden dense warmth on his skin through the window glass. Shelley's own warmth, the excited heat from her hungry red lips, brought her together with the sun in him. She was connecting him to everything, and connecting to everything in him.

But now, not three months later, everything was lost again—and the midsummer sun had somehow turned freezing. Henry gestured, showed me how he had covered his eyes—hands together, fingers interlocking—to block the sight of her driving away from the parking lot of the diner, where she had met to break up with him, just two days ago. He had been shielding himself from the image of her bright red rear taillights, as she left him to find another. Nothing remained for Henry—he had no connection to her, nor to anyone else, it seemed.

Henry's blockade of the scene of her departure seemed an oddly immature defense, I thought, more suited to a toddler than a grown man. He was midperformance, reenacting here in Room Eight—and watching me instead of his hands, closely tracking my reaction. As I looked on, and as he lifted his arms higher, the sleeves of his loose sweatshirt fell back to his elbows, exposing forearms crisscrossed in fresh razor slashes—crimson, crude, brutal parallelograms. A big reveal, an intended one it seemed, of agony and emptiness. His barren core was now visible, through his own shredded skin.

At that moment an image came together in my mind, labeled with a short diagnostic phrase. All the cryptic threads of his symptoms, each mysterious on its own, made sense because of their mutual intersection in that instant: the bloody violence of his thoughts about others, the cutting of his own flesh, his bizarre behavior on the county bus—and even the covering of his eyes to not see Shelley go.

The phrase was *borderline personality disorder* (a label of the moment in psychiatry that may change in time to something more reflecting symptoms, like *emotional dysregulation syndrome*—but which, regardless of label, describes something constant and universal, a fundamental part of the human heart). These three deceptively simple words clarified Henry's chaos for me, made some sense of his bewildering complexity, and in particular explained the positioning of his mind on the border between unreal and real, between unstable and stable. He was blocking the light's path to deflect the harshness of the knowledge it carried, protecting his raw and damaged depths, asserting crude control of what could flow into his body across the border of his skin.

Though each case is different, and I had never seen a person with a combination of symptoms quite like Henry's, new details began to fit the pattern as I asked more questions. He eventually disgorged again the fantasies of eating people with which he had shocked the paramedics—never actually harming others, but hating strangers on the street simply for being human. When he saw people he saw their insides, and their insides inside him.

The sun hurt, was cold and strong—and so to re-create the original feeling when Shelley had kissed him on the church shuttle, Henry had disrobed on a county bus, seemingly trying to find some patch of skin where the sun would feel the same. He was seeing blood everywhere, was swimming, diving, drowning. Good enough for police transport, state code 5150, to the nearest emergency room, to me.

Some of the people arriving on a 5150 hope to avoid an inpatient stay in the hospital, while others seek admission. My role was to make the border of the hospital real, by finding out who needed help to stay alive. My forced decision as the inpatient psychiatrist was binary: discharge Henry back into the evening or admit him on a legal hold—to our locked unit—for up to three days with no right to leave, an involuntary patient.

With the diagnosis now in mind, it was time to think about writing the note, completing my assessment, and settling on a plan—and that meant beginning with his first words. I looked down at my notes and returned to the moment I had walked into Henry's life.

•

Before money from our latest tech boom had flooded the region and brought about the emergency department's modernization, tiny Room Eight had served for more than twenty years in the valley as a major portal for incoming acute psychiatric patients. Many of the individuals who designed and created our densely connected silicon world had passed through this isolated latrine-sized room at one time or another. The valley was their home, and this their hospital, and windowless Room Eight served as the portal to acute mental health care—and thus as a sort of window into the valley's most human, most vulnerable heart. Room Eight was important; in a home it matters what can be seen from the window.

But Room Eight was dim and cramped, just big enough for the patient's gurney. Outside, an amiable, blazered guard stood by. Inside, a single chair for the psychiatrist was positioned as close to the door as possible; the ER setting can be unpredictable, and emergency psychiatrists (like other acute-care medical specialists) are taught to identify flight paths for themselves, and to position themselves close to escape routes, in case the interaction goes awry.

On my first contact with Henry, planning a flight path had seemed relevant. In a baseball cap and jeans, Henry was taller and heavier than me, unathletic but muscular—and his face seemed to twist loathingly at the sight of me. I tried to keep my face impassive, but my abdomen felt knotted and drawn tight in response. I had left the door cracked open, and as I introduced myself, sat down, and asked what had brought him in, the familiar cacophony of the ER filtered through: accompaniment for the first words of his monologue, which as my medical training dictated, would have to constitute the opening line of my note.

Psychiatrists begin as doctors of the whole body, in emergency rooms and on general medicine units, diagnosing diseases of all the organ systems, treating illnesses from pancreatitis to heart attacks to cancer, before specializing in the brain. In this yearlong all-purpose phase of internship after conferral of the MD degree, medical rituals are consolidated—including the rhythms of how to pass along all the information about a patient, in exactly the order expected by the attending physician (the senior doctor to whom the case is presented). This canonical sequence begins with the trinity of age, sex, and of course chief complaint, or chief concern—the reason given by the patient, in the patient's own words, for showing up in the emergency room that day. The formulation of *seventy-eight-year-old woman, chief complaint of worsening cough for two weeks,* is stated before anything else, before medical history, physical examination, or lab tests. This ritual makes sense in medicine, establishing focus on the active issue in a way that is helpful—especially for patients with many chronic conditions that would together otherwise be a distraction.

But medical custom is not always translated easily to the reality of psychiatry, especially in the next year of specialty training following medical internship. It takes a little time for the newly fledged residents, now in a phase of resetting and relearning, to transpose this medical rhythm into the new space, since the first thing the psychiatric patient says, when asked, can be awkward to restate as the first line of a medical note: *twenty-two-year-old man, chief complaint: "I can feel your energies in me"*; *sixty-two-year-old woman, chief complaint: "I need Xanax to cry in therapy"*; *forty-four-year-old man, chief complaint: "These fucks trying to control me. You can't follow me in death now, can you. Fuck you."* We write it down anyway.

I had elicited Henry's chief complaint with my stock opener, asking what had brought him here to the emergency room—and conscientiously recorded his response, the first line of my note:

Nineteen-year-old man brought in by police, chief complaint: "My father said, 'If you kill yourself, don't do it here at home. Your mother would blame me.'"

I recalled having so many immediate questions at that moment, but no pause had been given by Henry—he was only getting going, opening the veins. The words had flowed quickly, in a fluid and organized way—and everything, in retrospect, fit the borderline diagnosis. He implicated that broken relationship as the root cause of his suicidal despair, the lost perfect love that had begun just a few months ago with a kiss on a church field trip, and ended two days ago with their breakup at the diner in Santa Rosa. He had recounted from there the rest of the abbreviated, tortured odyssey occupying his past two days—learning to cut himself in secret, going to his father's house to show him the results, and after his father's stunning statement, running out the door and down the street, desperately searching for a bus, in a frenzy to feel what he had felt first with Shelley. Along the way, Henry included the story of his parents divorcing when he was three, complete with memories of climbing up on his mother's lap, crying *don't want to get that new dad*—but her face had been composed and set, impassive, comfortable with her son's tears. He had described the chaos of the divided home that resulted, when those who most loved each other became overnight those who most hated each other. How all human values, positive and negative, had been inverted, inexplicably, inescapably. How he learned to live with two separate worlds in two houses that could never interact, how he could not speak of one to the other, how he was forced to create and maintain two distinct and incompatible realities to survive.

And finally before he fell silent, he entrusted me with the visions he had described to the paramedics and the ER staff—the images of blood and cannibalism, and his revulsion for other people. Not just a desire for distance, but a disgust with all humanity.

Earlier as a medical student, I might have misdiagnosed him with schizophrenia or psychotic depression—dislodged from the real world either way. But Henry was lucid, and his thoughts were organized; he had not quite broken away. Only the person with borderline can travel from reality to distortion and back, speaking both tongues with dual citizenship—not quite delusional, but with an alternative frame-work—to help manage a hostile, unpredictable reality.

Sometimes it can seem that both the self and what lies outside the self are not yet fully defined within the minds of borderline patients—not well resolved as entities with constant properties and worth. The relative values of different situations in the world, and of the different levels of human interaction, seem not smoothly compared—leading to reactions without subtlety, such as catastrophic thinking about unlikely possibilities, or extreme reactions to the natural give-and-take of human relationships. It is as if they are still in an early stage of developing a kind of currency exchange that allows human values in different categories to be fairly compared—and so to guide feelings, and actions, in measured ways.

But this pattern of extreme and seemingly unwarranted reactions (which can also be present in other conditions, and occasionally appear in anyone) also seems to constitute a practical strategy for surviving the early-life trauma that so many borderline patients have suffered, a reflection of their reality that there is not a single or consistent value system that makes sense in the world. And other aspects of personal development can seem frozen in early states as well, such as the use well into adulthood of transitional objects like blankets or stuffed animals, items that soothe a child when they are held tight, allowing the security of one environment to be made portable into an insecure space. Henry's shielding himself from the sight of Shelley's departure— this was the defense of a child, in blocking rather than addressing an unbearable, unacceptable reality. All of these behaviors can be unsettling to friends and family and caregivers—but with reflection, and with experience, also can stir compassion.

Many borderline patients (and those who are not patients, but who still live with some of these symptoms) manage to keep private this fragility: the sudden swings, over an aching emptiness. Some are guarded about a secret curse too, one that is also a silent deliverance: intentional opening of their own skin, the volitional cutting of the arms, legs, and abdomen. These are wounds that need never be shown—except when useful. What need was fulfilled for Henry here, with the seemingly deliberate exhibition of his cut skin? Did he reveal this knowing what would be triggered in the system—in my system, in me? Borderline patients can seem maestros of eliciting emotion, bringing forth overwhelming negative or positive feelings—approaching their own in intensity, but within others. This skill can bring desired outcomes, rewards of a sort, including admission to the hospital (sometimes the underlying goal, even where there is no suicidal intent).

The more I now thought about the timing of Henry's gesture with his arms, performed while clearly watching for my reaction, the more it seemed manipulative in the moment, a power grab. He was not actually a suicide risk, I thought (my thinking swayed by the demonstrative nature of the gesture), nor did he hallucinate blood, nor did he want to eat people. Nor was he obviously criminal or antisocial; as far as my history-taking had revealed, Henry had never harmed a human being besides himself—not even an animal. And since he had never actually tried to kill himself, I assured myself that Henry probably didn't want to die, at least not yet. Though his pain was real, the showing of his self-harm was something else, a frantic grasping through borders real and unreal to find care and human connection, reaching across his own skin into others, diving under and clutching frantically at the warm blanket of human interaction that could turn cold at any moment, seeking the deep bond that would never come again. Skin on skin. His mother's face set, impassive.

I had pressing clinical issues going on—active players on the consult-liaison service, transfers from outside hospitals coming in, and a possible gastrointestinal bleed brewing on the locked unit. My capacity was not infinite. Henry suspected this, perhaps, and was telling his story strategically, knowing that if he did it right I could not easily send him out that night, alone, into that cold Palo Alto floodplain. He just wanted something from me, something immeasurably precious: me—my time and energy.

As this realization landed, I felt a rising tingle up my back, in that sensation of defensive rage that we feel in our skin, when personal boundaries are violated. Even though I knew his pain was real, my compassion was to that point only clinical and intellectual. A deep and shared ancestral state now arose in me that cared nothing for my compassion. The hairs were rising up, past my neck to my scalp, in that ancient and furious and privileged experience of mammals—a feeling that defines our skins, our barriers, and our selves.

•

Each emotion has a physical quality, like the bubbly thoracic sensation of falling in love. The rage of territorial invasion is felt in our physical boundary, in the skin. In our ancestors this feeling may have arisen as a posturing, the display of erected hairs to increase apparent size, but now, for nearly naked humans like us, this feeling serves only internally, as an unseen legacy felt personally, for our own use within. And Henry had evoked it in me, he had reached into me, eliciting the same sense our forebears had felt a hundred million years ago, as soon as there was hair to erect. The skin organs along the neck squeeze sockets holding hairs, the hairs stand, the body grows, the shape presented to the world expands—this is me; there is more of me, you should know. I matter more. I am more.

That feeling—nameless, universal, compelling—is an entangled inner state of positive and negative, an exquisite tingle of pleasure and rage. Elevated and expanded, my perspective grows and I feel myself rising too—a surge as the hairs stand. I am emboldened; danger is now to be sought—risk is everything; in this moment I can face the consequences and carry them through to wherever they take me. The boundary is the feeling, and the feeling the boundary. And then the hairs on my neck and back slowly lie down; I have a medical license; I am a professional being wearing a white coat, on a civilized planet, with limits. The wave crests and recedes. The feeling, with its primal mastery, subsides.

I had felt this before with borderline patients, but perhaps Henry could not know that he was bringing it forth now. Babies also motivate strong feelings in parents, without ever being taught to do so. Henry was young and unschooled, a baby borderline. He was a human mammal from a broken burrow—burrow broken at three and born then as borderline—frozen in time, with childlike defenses, yet with tools at the ready to breach my own boundaries, to move across my border, get under my skin, and tap my resources all the way down to my deepest, oldest inner state.

The skin is both border and sentry. Skin arises from ectoderm, in embryos; ectoderm is our initial boundary, the surface layer of cells, creating the most fundamental borderline between self and nonself. Our sense of feeling, our watchtower at the border between self and world, is built from ectoderm, with skin-embedded organs that detect touch, vibration, temperature, pressure, and pain. And the brain itself, though internal today, is built from ectoderm too, and so that layer ends up setting all boundaries of the individual, psychologically as well as physically.

Hair and fur are also built from skin, and likely came from whiskers first—muzzle fibers for touch sensation in our oldest burrowing ancestors, hiding from the surface-dwelling dinosaurs for forty million years until a meteor strike upended everything, sending mammals to the emptying surface sixty-five million years ago as most other life headed to extinction. These earliest hairs sensed the shape of the burrow in darkness, the dimensions of passage for the head—assessing if the self could enter for warmth or escape, designed to take a measure of earth's intimacy.

As whiskers evolved toward thicker and denser numbers, for increasingly rich sensation in navigating our dark burrows, we blundered into a new way of building borders. Thermal insulation with hair was discovered, and then ported around the body by natural selection in all its blind power. The burrowing mammals born with denser sensory whiskers also retained more vital energy—better managing their costly warm-blooded lifestyle fast-burning in the cold of night—and survived the sudden chill of a blocked-out sun.

These predesigned sensory skin organs then spread across the body over thousands of millennia—where yet further uses were discovered. Hairs could be raised upon threat along the neck and back, serving like a rattlesnake's warning; our earliest skin organs, like border sentries, were now also responding to invasion as a concept in the external world, as crossed territory, as new topology. And though raised hairs were an outward signal meant to warn others away, by the time we (mammals capable of describing inner feelings) had arisen, this visible sign accompanied something else harbored within. An internal sense was part of the state, becoming a signal more useful for the self. Hair—a mere peripheral skin organ far from the brain—was now reporting on personal territorial integrity that could be psychological as well as corporeal, and was signaling the invasion both to the world and back to us.

We (as humanity) would eventually lose most of our body hair once again, but the feeling itself remained, that exquisite charge of threat and growth—perhaps the first distinctive mammalian inner state, truly primal, a sense born long ago in dark tunnels.

We sense and define the borders of ourselves with skin: boundary, sentry, pigment, signal. Skin is where we are vulnerable, where our heat is lost, and where we must make contact to live and mate; skin plays many roles, and so bears its own diversities and contradictions. On our soft ventral sides, along the midline from throat to abdomen to pelvis—the front of a human being, derived from the ground-facing aspect of a four-legged reptile or early mammal—blood flows toward the surface for blushing and swelling, to reach out, to functionalize, to couple. But our hair-raising, tingling, rageful, boundary-violated feeling is instead felt and expressed dorsally, along the back—the more secret, less visible side of human beings, paradoxically facing away from the individual confronting us—but in our evolutionary history before standing upright, this was the more noticeable upper side, where like the ruffs and backs of cats and wolves, hairs could rise to help us expand our presence.

When the hairs are raised with rage—responding to loss of territorial integrity—some psychiatrists use that feeling in themselves, when they notice it, to help diagnose personality disorders like borderline. This clinical trick, rarely formalized, is an art of psychiatry, if not quite science—to listen to one's self, to notice the negative feelings evoked by the patient, to realize that those feelings are likely a shared response of others in the patient's life, and to use that insight in treatment. Thus an evolutionary vestige is also a diagnostic tool, remarkably—with all the caveats one might imagine, including being wrong—and so the wise physician maintains focus simply on the fact that the patient is likely to evoke that defensive feeling in others as well, which can be the source of difficulty in life, and thus the subject of useful therapy.

This transference works with positive feelings too. For good or ill, the patient or the psychiatrist might fit into a role from the past that had been created and played first by someone else in the other's life. By chance or desire, we sometimes find ourselves to be square pegs for square holes, and if the role is positive, the therapeutic connection can be strengthened—as long as the transference is identified, and monitored, and not allowed to distort the process of care. And indeed, almost inevitably in retrospect, toward the end Henry let slip a single sentence that helped me connect with him—unwittingly, or perhaps playing me perfectly. I had begun to wind down the interview, feeling more certain there was little risk he would harm himself that night, but still uncertain about admission or discharge, when he said, *I just want my parents to be together.*

There, right there, amid the tricks and misdirection, that at least was true. That was the one thing that mattered. The latent hope to connect the frayed edges and repair the broken self. A single father, I heard my son—and felt again, for a long moment, the fragmentation of our home when he was two.

Aware of the transference, and reminding myself that I could do little—and understood less—I admitted Henry on the 5150, completed the paperwork, called the unit, and brought him in to keep him warm.

•

Mostly unresponsive to medications, borderline is a perplexing brew of symptoms that can seem unrelated to one another: frantic fear of abandonment, intense mood swings, inescapable feelings of emptiness, bizarre public displays, morbid visions. Suicide is more common in borderline than in any other psychiatric disorder, and nonsuicidal self-injury—as with the deliberate cutting of flesh—can become powerful and rewarding, even desperately sought. This is a behavior that few can fairly claim to fully understand, but cutting is common—and thus means something about us, about humanity.

Unlike in other psychiatric illnesses—such as schizophrenia where bizarre symptoms force disengagement, push away others, and isolate the patient—borderline symptoms can often engage, entwine, and draw in others, at least for a time. Self-injurious actions like Henry's may indeed entangle others this way, but seem to serve some purpose for the patient too, internally. There is already another pain, of a different kind—and self-injury may work against this deeper, more profound hurt.

We know that many of these human beings have borne an unjust burden: psychological or physical trauma at a young age, sometimes attributable to their own caregivers. Henry's only source of warmth, in his tiny family nest deep in the chill of a redwood grove, had been not just disrupted, but also upended—inverted in value—and whatever had actually happened with his parents, Henry's perception and interpretation were clear: he had suffered greatly when he was very young. But care coupled with pain still adds up to survival, in a calculus of practicality for adapting to a hostile and confusing world. If those we trust, those we must trust, become unpredictable or harmful and borders are crossed, if value becomes inverted in a fundamental way, then a strange new logic for staying alive is needed. Survival requires staying engaged with caregivers, and not everything really needs to make sense if it also brings warmth. A ruptured world order results in a ruptured emotional life where nothing is stable but must be stabilized, and where human connection becomes a dialectic: both desperately needed and utterly to be avoided. In this light, the ability to work with alternative realities, in oneself and others, begins to make some sense.

The correlations of these diverse symptoms are real and can be quantified by epidemiologists. For borderline patients, trauma during dependency—early in life when warmth and care are needed at all costs—predicts nonsuicidal self-injury later. And the human dependency period is long. We have vast and intricate brains to build, and a diverse civilization to assimilate—a babelized complexity of human custom and cognition—which can best be done with the trust and speed and accepting nature of the child's brain. Our brains are building even basic structure—the electrical insulation, the myelin that gives white matter its whiteness, that guides electrical communication pathways across the brain—well into our twenties and beyond. As primates, and as human beings, we keep our skin exposed—dermal or neural, borders or brains—available, for use or abuse, as long as we can.

And so primate evolution in the direction of modern humans has brought us a dramatically longer childhood, with dependency and vulnerability greatly extended in time. Childhood is now pushed to its limits, longer already than the mean life span of our recent ancestors, to the borders of fertility and beyond. Nowhere is this phenomenon more clear than in the practice of medicine itself, with its interminable training period. The halls of teaching hospitals are populated with small clutches of doctors still in residency or fellowship training, schooling in tight and vulnerable clusters of white coats. They are all in mid-adulthood but still trying to learn, trying to find love, and trying to not die—their hair a graying outcropping of the skin and the self, signaling frailty more than authority.

Though we know why our vulnerable period may be so prolonged, we do not yet understand the biology of borderline personality, at the levels of cells or circuits. As always, to approach this question in a scientific way, we might choose to simplify by reducing the selected question to a reliable measurement, to a single observable. The reward of pain, of cutting—though not unique to borderline—is coupled to the disorder, and serves as a remarkably clear measurable, reporting on a powerful and altered human inner state.

What makes a human being cut? This is already a hard question, but one can take it a level deeper: what makes any being do anything? In different settings, the answer can be reflex, or instinct, or habit, or to avoid discomfort or pain, or to get a bit of pleasure, a buzz of reward . . . but instead, we could imagine a world where all behavior is guided by pain, and by release from pain. We sometimes are driven by the seeking of a positive feeling, but an individual can instead be guided chiefly by suppression of internal discomfort as the motivation for action.

Could behavior be motivated for survival strongly enough if a species or an individual were to do without pleasure—and instead just use temporary reduction in suffering as a motivation for the correct behavior? With a suitable action taken, to promote survival or reproduction, then an internal pain would be reduced—if only for a moment. If we were gods designing beings, this strategy might work. What would a human being look like, act like, whose baseline was psychic pain, and whose every action was taken to reduce, or distract from, this pain?

We can put off pleasure at any moment, but we cannot as readily ignore pain. Perhaps, then, pain would be an even more powerful force for guiding behavior. Reduction of, or distraction from, inner pain might work as a motivation for just getting up in the morning, or for socializing with friends, or for protecting children—though the mannerisms might look odd to us, as we are constructed now. The style and melody and rhythm of each action would seem off, bizarre and volatile, of a being living in agony and acting for its reduction. But such an existence, at least for some, may already be a reality. Such people might not seem too different from human borderline patients—our sisters and brothers and sons and daughters, terribly burdened with negative inner states.

Such an insight could also bring hope for understanding and treatment, because inner states and value systems can be changed, and may even be designed for easy change. As a being grows, as the environment changes, as the species adapts and evolves, valuations assigned to parts of the world—like the worth of having a thing or being in a place— must adapt as well. Such internal value is a currency like any other, and should not be fixed to an immutable standard that could prevent growth. Instead, value must be set by fiat—whatever is good for survival—and easily, precisely, quickly so. From birth onward, as the dimensions of self and life change, existential dangers—even threats to life, predators—become minor annoyances, or objects of beauty, or prey. The rush of fear and horror must fade, must become joy, must morph into the thrill of giving chase.

Changing of value on any timescale—fast with an instantaneous new insight, slower with growth and maturity, slower still even over millennia as the world and species evolve together—allows adaptation to shifting conditions by tuning inner exchange rates for the competing currencies of suffering and reward. The experiences of borderline patients, and insights from modern neuroscience, together show that valence—the sign of negative or positive experience, aversive or appetitive, bad or good—is designed to be changed, and readily.

•

Neuroscientists can now set these exchange rates, adjusting with precision how likely an animal is to do almost anything, with optogenetic targeting of specific cells and connections across the brain. For example, depending on the specific circuits targeted, we can cause animals to become more or less aggressive, defensive, social, sexual, hungry, thirsty, sleepy, or energetic—by writing in neural activity with optogenetics (in other words, dictating that just a few spikes of activity occur in a handful of defined cells or connections).

As the subject's behavior changes instantaneously, shifting to the favor of one pursuit over another, and thus seemingly switching from one system of values to another, sometimes a psychiatrist cannot help but think of borderline patients. These individuals can be swift to react strongly with value assignments or changes—for example, treating a new acquaintance or a new psychiatrist as an archetype of the category: the deepest friend, the best doctor. And this positive categorization so powerfully expressed can be erased or reversed in an instant—transitioning (after a caregiver's perceived misstep, or after attention from a partner is perceived as inadequate) from best to worst, all the way to catastrophically negative.

This binary switching in people is sometimes attributed to skilled acting and manipulative intent—but my perspective (shared by many) is that these labile states are truly felt, overwhelmingly so. Extreme reactions reflect all-or-nothing feelings, subjective states adapted to uncertain life experience. The survival skills of a traumatized child—though this does not describe all patients with borderline—become the distortions of a suffering adult human being, living life in chronic negativity, with everything framed in terms of what might be strong or pure enough to distract from unrelenting sirens of psychic pain broadcast throughout the patient's inner world.

There are deep and powerful brain structures through which these effects can be borne out. Some of these circuits and cells (like the dopamine cells near the brainstem) broadcast their influence far and wide, sending connections nearly everywhere in the brain—including to the recently evolved frontal regions where our most integrative decision-making and complex cognitions occur, as well as to older regions that manifest survival drives in their most basic forms. Positive or negative value can be easily attached by these dopamine cells to even neutral items like an unremarkable room. With optogenetics, turning down the electrical activity of dopamine neurons in the midbrain, by providing a flash of light every time a mouse enters a neutral room, causes the animal to begin to avoid the harmless room, as if it were a source of intense suffering.

This experiment may be accessing a natural process, since a different but interconnected deep brain structure, the habenula, (a structure—so ancient it is shared with fish—that fires away during hopeless, uncontrollably negative, and disappointing situations), acts through turning down the dopamine neurons in the midbrain naturally, just as optogenetics does experimentally. This circuitry can thus impose sign, or valence, where none was present before.

It has been discovered that early-life stress and helplessness can increase habenula activity, and borderline patients may be locked in constant uncontrolled negativity from their habenula-to-dopamine neuronal connection—or other related circuitry. Fixed to a baseline of pain, they may live out a hard-learned lesson about the way the world is that could have only been internalized by the young.

Cutting may reveal such a negativity of the borderline patient's inner state. This behavior might recalibrate that negativity, introducing a new, sharp, and fresh pain that is controlled and understood, rather than the uncontrolled (and inexplicable) feeling from childhood. So lifelong suffering, at least for a moment, is renormalized into almost nothing by comparison with the new self-generated sensation. Intense negativity—as long as it comes with agency, with control, with a reason—can be desperately sought.

Modern neuroscience may thus begin to reveal how Henry, and those like him, could come to dwell in such a state, with early-life trauma seeding negative-valence predisposition into the arable field of the young and vulnerable mind, and sowing deep instability in the valuation of human connection. Studying fish and mice, our cousins with whom we share key ancestors, shows how powerfully and instantaneously the value of absolutes can be accessed and changed by activity in a few specific cells and circuits in the vertebrate brain—and thus very likely in our brains.

We each have a narrative in our minds, an in-progress drawing easeled up and ready to go to explain ourselves and others, to justify our sense of self and our relationship to the moment. We carry that depiction around with us, and we also carry those of our friends and family and others who are important to us, as images which we consult from time to time. It has been hard for those who most love and cherish the borderline patient to build such an image—to really create and hold an internal model mirroring their loved one's narrative and suffering.

But with a little help from modern neuroscience, these friends and family and caregivers and others can now begin to imagine, and perhaps nearly understand, living life this way.

Early-life trauma can happen to any animal, but our young may be most vulnerable because they have the most to internalize. Our evolutionary (and cultural) strategy for learning has been to lengthen childhood, and so as a side effect, extend risk. Other animals might for other reasons come to live in negativity as well, without a means or reason for signaling this inner state to the outside world, but borderline-like symptoms may most readily reveal themselves in the context of the complex social networks of human life—and when our unique planning and toolmaking allow for discovery of behaviors like cutting. Even Henry, as I found out later, did not stumble across that particular innovation on his own.

•

Henry had many superficial cuts on his arms that were rapidly healing and uncomplicated. As borderlines go, he was still mild, still just figuring it out. Even his known childhood trauma was not so bad, at least as known to me, at least by comparison with what I had seen in others—a difficult divorce for sure, but far worse can happen.

And yet Henry's suffering was real. His family was broken, and every experience that he shared was warped in some way by this fundamental loss, which was a burden squirreled away whole, unmeted, bending his inner form, creating counter-confusions of positive and negative, black and white, reality and imagination, until the only dialectic that mattered was the one at the heart of everything for him: connection and abandonment, water and oil, unblendable.

For the first three days of our hold, a measured process of ministration was set in motion—of set rhythm and duration as with any 5150. The one held, the newcomer, is made warm and then made available, like a new lion cub introduced to the pride; the patient is first given a bed, and then in firm and steady ritual is visited by members of the caregiving team. Several days follow of this gentle and insistent attention—from nursing assistant, nurse, medical student, resident, physical and occupational therapists, clinical psychologist, medical consult team, social worker, attending physician — alongside the other patients, all strangers held together and each with a different reason for landing there. It is altogether a more complex and challenging brood than could be prepared for by instinct or intuition.

The time spent by any one patient on a locked ward is typically just a few days, which seems not enough time for cells or circuits to change fundamentally, nor for significant therapy-driven behavioral modifications. Yet each morning, a life-or-death judgment must be made by the clinical team on the locked ward. As we evaluate patients subject to the 5150, those truly recovering and those simply recanting cannot be easily distinguished. All we have to make these judgments are human interactions and words, alongside published statistics and accumulated individual clinical experience. This is not enough; still, deep in danger we estimate the risk, because there is nothing else to do and there is nobody who knows more. Each day we must decide to continue, or release, our hold.

Even more unsettling, a deadline bears down. By morning of the third day, the hold expires and the patient is automatically released into the world even if danger continues — unless additional steps are taken. Numerology seems the only relevant consideration in setting the term of this three-day limit, since the duration maps onto no specific medical or psychiatric process. Three days, compelling and biblical, Old Testament or New: *three days and nights in the belly of the beast, three days and nights in the heart of the earth.*

If acute suicidality continues, two more weeks of care can be sought in a different kind of hold called a 5250 in the California code. But then judgment truly comes calling, in the form of a nonmedical outsider with right of passage into the psychiatrist's territory. This is the hearing officer, a judge arriving on the unit — trailed by another visitor, the "patient advocate," who is to play the role of pushing for release. The doctor (if still feeling that release may be unsafe) can make a plaintive case to continue care, to keep the hold — only now, this happens against opposition. This is an uncomfortable charade, a doctor arguing against someone named the patient advocate — while the doctor's whole calling and sense of self is built around helping patients heal in safety. Yet doctor and patient advocate must rise in battle, civil and gracious yet with secret ruffs half-risen, necks itching.

When animals within a species come into conflict, natural mechanisms from ancient circuitry can act to minimize damage. Rituals signaling size (as with measuring wide-gaped mouths against each other, in hippos and lizards) often allow the smaller rival to escape safely, and both to conserve energy. This conflict avoidance works when the stakes are not life or death, as in many mating conflicts where other such chances are present or may come later; but if opportunities are sparse, de-escalation of the conflict is harder. In the hearing on a locked unit, in these rituals, no de-escalation is possible, and the stakes are existential—truly life or death, but not for the combatants. The one with life interest in the outcome, the patient, waits in another room, with no presence or voice.

I had won almost every hearing before this one, and expected the same with Henry. But after only a few minutes, the ruling from the hearing officer, coming with godlike finality, was that I had lost. The edict for Henry was discharge: freedom and danger.

With no personal stake in the decision, I should have found letting go to be easy; but the outcome of this one was hard for me, and I found myself running through the case and the hearing in my mind again and again. Objectively I could understand the hearing officer's decision. Though I was concerned that Henry had not contracted for safety—he refused to make a promise to not seek suicide—his self-harm so far had been undeniably nonlethal. That fact was enough for the hearing officer, and perhaps it should have been enough for me.

I also should have been pleased that such a high value had been placed on personal autonomy by this decision, since I valued freedom too. I understood—all sides understood—that if a secret suicide were planned, it could now proceed, but in this case personal freedom had been deemed of greater significance than that small risk, upon the balancing of two categorically different fundamental values. This subtext is the core conflict of every such hearing—patient freedom versus patient safety—and thus both sides are the patient advocates in a real sense. Advocates for autonomy or for security: there is no older or deeper conflict, and none closer to the beating heart of borderline.

I struggled with this verdict, but I understood a source of my internal conflict; I was not blind to the transference. The parallel with my own life was not subtle—at least in one aspect, the early-life collapse of the home—and I could not help but wonder about my own son, only five years old at the time I treated Henry. While no hint of his affliction ever showed in my son, I did not have that perspective on the day of the hearing—and Henry developed his symptoms late. It was not until after his summer breakup, nineteen and with the sun still cold on his skin, that he had watched a movie on his laptop showing explicit cutting in a girl of thirteen—and the concept had clicked, hard, for him. He tried it right away, copycutting with crude tools and coarse strokes behind the community college gym, and then went straight to show his father.

Why did he go first to his father, to reveal his wounds? Perhaps it was simply to make the hurt known, to connect by shock and blood. But why not begin with his mother instead? She was the one he had seemed to fault at first: he had pointed to her as the one who had left the family, the one who had abandoned the nest. Henry's chief complaint—"*My father said, 'If you kill yourself, don't do it here at home. Your mother would blame me'*"—was this the key clue, a meaningful signature of pathology in his father instead, in ways that we did not yet understand?

These are the mysteries that a few days cannot uncover; obscured still, Henry's story had not been truly told. There was no time for deep connection. Henry had somehow revealed little of importance, that we could grasp anyway, in his two and a half days on the unit. He did ex-hibit a superficial form of progress, a decrescendo of sorts—gradually toning down his violent language, the descriptions of his desire to die, to drown in blood. But I knew how readily he could present different stories at different times, depending on what was needed, and I was not reassured. I wanted the time to help him.

If I had played the hearing differently, I might have found some way. Holds in California can be placed or extended, not just for suicidality, but also for danger to others, and for grave disability. But despite Henry's rage he was not violent, and never had been, to others; his bloody visions were just that: a churning of violent imagery not coupled to the power stroke of action. This left only the route of finding evidence for grave disability; perhaps a plausible argument could be imagined from his nakedness on the bus, a case to be made for his inability to provide for at least one of the basic-needs triad: food, clothing, and shelter. But Henry undeniably had, and knew how to access, resources addressing all his needs. The bus incident, like his cutting, was serious but not lethal, and so Henry stepped out of the unit on a foggy Sunday morning.

I watched him walk down the hall toward the escalator and the main hospital exit, canvas bag over his shoulder. He was uncured, untreated even, but I told myself there was little more that could have been done. His was a disorder untouchable by medication, he had wanted to leave soon after admission, and upon discharge he had refused even my out-patient referral for a specialized group behavioral therapy. The clinical literature predicted Henry's future would include more of these para-suicidal actions like cutting, that were revenging and rewarding in ways I would never fully understand. His wounds would heal and then reap-pear, as relief continued to come from the act for him—a desired in-jury, a counterstrike against internal suffering beyond my imagination. Henry had no choice; for a time he would have to continue to seek these stigmata, and to seek others—not skin to skin, but self to self, coercing human warmth across space and time.

His destiny in the long run could be the mellowing of borderline symptoms that usually comes with age—but time could instead bring suicide, the ending of the self, at a rate of 15 percent: the highest inci-dence for any disease, any burden of humanity. One hope was that those who cared for him could learn to use the state he was able to evoke in them, projecting that ancient feeling of the invaded self, mag-nified a hundredfold, back onto their own internal representations of Henry. Powerful empathy can be stoked from sparks of anger.

My own flare of rage had long since died away, though I knew I was still vulnerable to him, and would always be. Henry projected into me, and was close to me, as the written word is close to paper. But I felt that

I had exhibited to him only a pleasing dupability, in my earnestness to reduce his pain. And for a time I couldn't see my son without thinking of Henry. He had written his story atop my own, like a medieval monk inscribing new text upon a scraped and reused parchment, carving symbols of judgment and revelation onto animal skin stretched thin.

CHAPTER 5

THE FARADAY CAGE

Hegel made famous his aphorism that all the rational is real and all the real rational; but there are many of us who, unconvinced by Hegel, continue to believe that the real, the really real, is irrational, that reason builds upon irrationalities. Hegel, a great framer of definitions, attempted with definitions to reconstruct the universe, like that artillery sergeant who said that cannon were made by taking a hole and enclosing it with steel.

—MIGUEL DE UNAMUNO, *Del sentimiento trágico de la vida*, translated by J. E. Crawford Flitch

The new thoughts came with all the surety of a change in season, in a gathering together of signs. Like the air of early fall, the first few weeks seemed to bring a shift in pressure in her mind, with a hint of wind revealing itself—a shimmering of her highest leaves, a rustling in the neural canopy.

She could feel the change in her skin as well, a subtle tingle, a chill of early fall. The sensation stirred a memory from a dozen years ago: Wisconsin in September, with her brothers AJ and Nelson, chasing Canada geese along the lakeside. Winnie had been seventeen after that summer of lymphoma chemo. Nothing had ever felt so charged as her return to the outdoors that fall after methotrexate—all around her and within her, even to her lungs, and to her brain, a mist of the season had seemed infused, clear and crystalline. In remission they had said, a likely cure, and they were right.

But this time, with the rustling of the leaves, unsettling intimations had come, borne high and kitelike on the same ghost wind—and there was a feeling of openness, of vulnerability, that was not entirely positive. She abruptly decided to take a month off, which was unprecedented for someone with her caseload. The team muttered, including her supervisor, but Winnie had built serious credibility, even a kind of celebrity, winning allowance after allowance, constructing patent estates from chaos—wielding her mind like a weapon trained in both law and engineering, unique in its ability to grapple with interlocking artificial-intelligence intellectual property families. Her team of lawyers and staff had filed seventeen hundred patents—counting all the divisionals and continuations—for their major client last year alone. But now she needed a month of leave; there were pressing issues to address. She was exposed.

The first issue was Oscar, who lived in the townhouse next door. He had installed a satellite dish on the roof over his deck—and seemed to be preparing to download her thoughts. Winnie needed someone to pay him a visit, dismantle the dish, and take him into custody; her homeowner's association security would have been natural to call in, but they were probably allied with him. Same with the police. She needed to find a do-it-yourself solution, to take care of herself, as she always had.

A trick occurred to her, a temporary countermeasure against the satellite dish—just a quick kludge, but with a chance of really working. She dug out a heavy black knit cap, the one with the reflective Raiders logo she got in college, which she hadn't worn since her Berkeley days. She put it on and pulled it down tight over her ears. Right away, everything seemed more contained. It was a bit surprising that it worked that well, with just the silvery logo of a football team as an electromagnetic field insulator, but there was no doubt—the satellite signal felt less likely to get in, or her own thoughts to leak out. The hat's tightness helped to shape the air around her head, to separate and clarify borders.

This vulnerability was correctable, then, and a more permanent solution dwelt in engineering. There were structural changes she could make inside the bedroom wall to reinforce that border with a conductive material—installing a true modern Faraday cage as a shield against the satellite dish signal. She started working on the wall, and her home tool kit gradually expanded with a few short drives to the hardware store across town for some more specialized items—a crowbar, a little chicken wire, some sheet metal, a voltmeter.

But other developments in this strange new season were more disturbing, and harder to address. Outside her expertise. More biological. At the center of it all was Erin, assistant to Larry the senior partner, younger than Winnie and five months along. Erin had become pregnant clearly to taunt her, targeting Winnie for living alone and not having children. It was unprofessional of her, and embarrassing for Winnie, and a little frightening considering Erin's proximity to power in the firm.

Here, there was no clear engineering solution to address the offensive behavior. Winnie had to get to Larry himself. Larry was the only one who could discipline Erin, and he needed to be informed and challenged to act. So over a weekend Winnie planned an incursion into the upper reaches of her own law firm—the C-class floor, with all the chiefs and captains. She mapped out her access plan and rehearsed the conversation with Larry—mostly in her head at first, not using her computer or the Internet, as it could be assumed Erin had hacked everything of hers, and had long since obtained access to her emails.

A lot of her planning became sketching on paper, elaborate reconstructions of desk orientations and restroom placements from memory, but then she got restless, needing to move her body, to do something physical, and so Winnie went back to her satellite dish countermeasures for several days—removing drywall, peeling the insulation back from her east-facing wall to see what lay behind it, and beginning to arrange the new metallic shielding.

Then came new, darker undernotes to the change in season, and some were frankly frightening. Over the second weekend of her leave, she became aware of the grim and gray-lipped ones—information vampires. Stocky, thick, and strong as ox hearts, lurking in the shadows behind the dumpsters, they started draining her energy and thoughts, tapping directly into her. And with this she moved beyond, into some new phase. This new season was not just wind shimmering her leaves. No longer just gentle phantom fingers softly stroking, now it felt more like glumpy digits pinching coarsely, aggressively, for her cells like grains—her brainpan a helpless, squat saltcellar.

And then, finally, a new voice emerged on Sunday, from within her head—mid-pitched, ambiguously sexed—intermittently repeating the word *disconnect*. The voice felt familiar in some way, with a quality she recognized from adolescence, when she once heard her own thoughts at that pitch, except now much louder and clearer. It was alien yet deep within her somehow, a shout between her temples.

On Monday morning, Winnie decided it was Erin, and knew that she had endured enough. She steeled herself, stepped out of her townhouse, and climbed into her car. The drive itself went smoothly, past the shadowy parking lot dumpsters without incident, though there was a surprising sharpness of the stop sign as she turned onto El Camino Real, standing out in a way that mattered. The acuity with which she felt its eight edges commanded attention—but a horn blared behind her. Startled, she drove on.

Winnie arrived ten minutes later at her law firm's oak-scattered Page Mill Road campus. She disembarked carefully. In the parking lot, near her car, a flattened screw lay on the concrete. She knew, as soon as she saw it, that they had placed it there as a sign: they knew she was coming and intended to screw her.

The day abruptly darkened—the atmosphere now ominous—and she nearly turned around and headed back to her car. An unsettling thought came to her: the screw revealed their deep access to her plans, since they knew she would be there, and therefore they had so much more—her personal life, her private records, her healthcare even. And she had gone through a miscarriage just a few days ago . . . though as she thought about that, her blood rising, Winnie felt her grip on her own experience loosen. She became not completely, not 100 percent certain there had been a miscarriage. She could not picture the experience or any details; suddenly, it was a little hard to remember what actually had happened . . . as if that inner wind whipping up to a tornado had stripped her branches nearly bare, and her memories were now mostly lost to that gray-fingered twister, swirling down from a glooming cloud that was heavy and pregnant with rain.

Winnie paused, trembling, standing over the screw, pressing her temples to process it all, to focus herself, to consider all the ramifications and uncertainties. A paralegal she knew distantly—Dennis, something like that, datable she had thought once—walked past on his way to the main building. He sent an odd look her way, searching. She turned away, put her sunglasses back on, and tugged her Raiders cap down tight.

Before others, anyone—lawyers, admins, paralegals—can complicate things, you have to go in now, she told herself, speaking the words clearly and distinctly in her mind as if lecturing. *You will not back down and run. That little screw message is just from Erin. Larry can be turned; Larry will be on your side.*

She steadied herself and stepped deliberately inside, keeping as far from the walls as possible; with a tense smile she showed her badge to the security guard at the desk, then walked to the elevators and rode up to Larry's fourth-floor domain. She made a pass by his suite; careful to avoid eye contact, she was still able to mark Erin at the admin desk, looking for trouble. Winnie's first tactical mission had been perfectly accomplished—identifying what Erin was wearing, that formless yellow dress. Then she headed to the bathroom, entered a stall, closed the door, and waited, positioning her line of sight so she could watch through the crack in the stall door for Erin to come in, knowing it could not be long.

She waited almost an hour, but eventually a flash of yellow flickered in. Winnie calmly stood up, opening the door to her stall just after Erin's closed. She walked straight to the bathroom door, exited left, pulled her hat down tight, and strode back to the office suite.

She'd worked with Larry on a couple of the firm's thorny international cases, but only at a distance. They were two different kinds of human being—diplomat and introvert, schmoozer and quant—and yet today he'd recognize her, and realize the urgency once she began to speak. She walked past Erin's empty desk, knocked on Larry's closed door, and walked in. He looked up from his laptop and made eye contact. She sat down, confidently, in a chair before his desk.

There followed a great deal of confusion. It could hardly have gone worse, and it seemed the next moment she was sitting in a cluttered side room in the human resources offices, waiting for an ambulance, with what must have been the entire firm security team, perspiring in blazers, watching her.

She had been forceful but scrupulously polite with Larry, factually laying out the situation—describing how Erin's pregnancy was an unprofessional gesture planned to humiliate her, providing details of the email hacking that had transpired since, even telling him about the screw and how terrifying it was to see—but she had also maintained, she thought, a reasonable and calm tone. She had kept her face expressionless, solid as concrete—careful to not upset him with emotion or gestures—but it all had seemed to spin sideways after a few minutes. Larry was on the phone, then the first blazer came, then firm hands on her elbows. With her vision darkening in humiliation, she had been marched out right in front of Erin. Winnie made sure to give no route in, keeping her face a mask, with no eye contact—and off they went, to this windowless room she had never known existed.

The ambulance came a few minutes later. Two men in purple latex gloves appeared with paperwork and a scattering of tubes and wires. She was relieved to see them, and desperate to get out of the little room. The EMTs were both thin and densely muscular like climbers—and courteous, as they conducted a quick physical exam and asked about her psychiatric history. She told them the truth: there was no mental illness in her family. But AJ, her older brother, was different from the rest, with a way of saying odd, muddled, arresting things. He had never found his path—but he also never really got the chance. Winnie told the EMTs how AJ had been found on a downtown plaza, collapsed alone near a bus stop on a blazing hot day, already dead.

It had been an AVM—arteriovenous malformation—an artery misdirected, its thickly muscled walls jetting high-pressure blood directly into a delicate vein that evolution had designed for another job: only for collecting puddles of spent blood, weakly exuding from the brain. A doctor had said the malformation might have been a sign of a broader problem, a connective tissue disease—but nobody ever knew for sure, just that at least one AVM had always been there, hidden from view deep inside his brain, struggling for years to cope with the ferocious and incessant pounding of the carotid pulse, its diaphanous membrane stretched thin until the moment came to burst.

She did also mention her possible miscarriage from a few days ago — she was still unsure about it, with the memory poised between real and unreal. They seemed irritated by that uncertainty, which she understood; she was bothered too. She was certain about her distant cancer, the words so familiar and fraught, lancinating even now — *cutaneous large T cell lymphoma with central nervous system involvement*. She recounted the clinical course expertly. How it had started with double vision and headaches . . . and how since they had found some cancer cells in her cerebrospinal fluid, the methotrexate had been infused directly there, into the spinal canal itself, at the level of her lower back. How she had been completely cured, now going on twelve years cancer-free.

She had some scrapes on her knuckles from the wall teardown she was doing at home, but she explained that only briefly since they didn't seem to care much about her renovations. She noticed the EMTs persisted for some time in asking about drugs, in every which way, perhaps trying to trap her, but getting the same answer again and again — no drugs, never even a cigarette, just an occasional glass of wine. In the ambulance, things finally got quieter, and she had a little more time to think about it all — a frustrating interlocking puzzle of possibilities.

Most likely her thoughts had been tapped, her plans picked up and sent ahead to Larry and his team by data vampires. Meanwhile she noticed the paramedics were calling ahead — to the hospital, they said, but more likely to the grim and gray-lipped ones. *"On a fifty one fifty"* they kept saying — 50-1-50, or 50-150, or 51-50, which was it? The code must matter. Was it used to trigger, or accelerate, a download? Normally she could crack this kind of thing. Winnie pulled her hat down more tightly, and tried to drift back in time, just a few weeks, to how it all started, to feel that first exhilarating breath of fresh September air.

Later in the emergency room, the nurses and doctors asked all the same questions as the purple-gloved gentlemen. They pretended to type her identical responses on various tablets, apparently never bothering to talk to each other, in between rounds of prodding and poking with stethoscopes, blood draw needles, and reflex hammers.

They didn't care about her renovations either, but were very interested in the story of AJ—oddly far more than the paramedics had been. It got hard for Winnie to talk about him by the fourth or fifth time. As she gave shorter versions of his story every time, longer versions came to her, within her. There were increasingly lengthy pauses—she would stop midsentence, even midword—as images came sweeping through. Imagined scenes of his final moments, alone without a sister there to hold him, with nobody who loved him to cradle his muddled head.

AJ—the lost child, lost long before he died. School was as hard for him as it was perfectly suited to Winnie and Nelson, right down to their precision handwriting and their love of logic and engineering. But for AJ, even odd jobs had never quite worked out, whether in car shops or bakeries. Every venture seemed to end with befuddled bad luck, poor judgment, or dumbfounding accidents—though he stayed gentle all the way, until he fell that blazing summer day. She flew back east for the funeral, and a wrenching bark of a sob burst from her own body, a sound she had never spoken or known, when she saw that familiar crease in his forehead smoothed out, at rest, at last.

She lay curled on her side, on the gurney in Room Eight, and became lost in imagining AJ's final moments, reliving his run from the bakery to the bank, the run she and Nelson had reconstructed from scraps of paper in his pocket and clues from his co-workers, the agitated dash that turned out to be his last desperate attempt to support an independent life. The doctors had said the stress of that day, the running, the heat, and the worry, all had probably brought his blood pressure up, and the AVM had finally ruptured. Just a frailty waiting quietly, a small thing knocked loose by a day in which all the things that made his life hard had come together at once.

They could still poke and bleed and scan, but Winnie was done. Day turned to evening, dry sandwiches and juice boxes appeared and disappeared . . . and then a long stretch of nothing.

.

A knock came at her door, and a doctor walked in, with disheveled brown hair and coffee-stained blue scrubs under his white coat. He introduced himself, seeming to be sort of a mumbler—or maybe he was just tired. Winnie didn't quite catch his half-swallowed name, but *psychiatrist*: that word she heard.

Winnie sat up and swung her legs over the side of the gurney. He shook her hand, and sat in the chair near the door, saying, "I've seen all the paperwork from the ER team, and I've spoken with the ER docs. But if it's all right, I'd like to hear from you, in your own words, how you ended up here today." Winnie looked him over carefully, and then she rested her gaze in his eyes, taking a moment to think about his angles, and hers, before responding.

At the end of the day, she needed help, and had found no allies yet. Best to tell him something, if not everything. "Information vampires," she said. He needed to know. He wrote it down, and looked right back at her. "All right," he said, "tell me about that."

So she did—well, most of it. Not every detail, just the hard facts as anyone would see them. The information vampires were tapping into her brain, draining her thoughts; that all was clear enough and she could be logical and calm in describing it, with a surfeit of evidence that she could tick through. First her neighbor had installed a dish antenna on his roof two weeks ago to gain better access to her thoughts, but she had a shielding countermeasure that was in process. She had stopped going to work since people at work were accessing her, hacking in, trying to decode her thoughts and feelings. She told him about the screw in the parking lot too, so he would understand how powerful her enemies were, and why she had to disconnect and protect herself.

Winnie briefly mentioned the *disconnect* voice—how it was frightening but also reasonable, speaking a word she might have thought herself, voicing an idea that she wanted, but maybe something an enemy of hers wanted too. She explained that the word was spoken audibly inside, with all the qualities of sound. Someone, probably Erin, was accessing her mind—but why, she did not really know.

After a while he began to ask his own questions, in a pattern different from that of either the paramedics or the other ER doctors. When he asked about the Raiders cap she was wearing—pulled down to her eyebrows—she told him plainly, "That's to protect my thoughts." When he pointed to her gurney and asked why she had pulled it away from the wall into the center of the room, she answered simply, "Because I don't know what's on the other side." He circled back to her renovations, which none of the other doctors had shown any interest in, asking her for the first time about the wall she was tearing down and why.

In the midst of the doctor's questions, though, his pager went off; he apologized and went away. She spent an hour alone, looking at the wall in front of her, and then he came back, rejoining without preamble, acting as if it had only been a minute. Winnie asked what was going on. "It was just an emergency on the floor, sorry. I'm almost done here, but I can tell you what's happening," he said, sitting back down. "We're waiting for a couple tests to come back, but the bottom line is nobody can find anything wrong in your body—every test and scan looks normal. So that means we think what is going on is psychiatric. And the good news about that outcome is that there are treatments that can help you."

Winnie was not surprised. It had looked increasingly like the ER staff thought things were headed that way, though it didn't really matter—at this point she didn't care what they said, all she wanted was to go home. The ER docs had informed she was "on a legal hold" and could not leave until psychiatry had seen her, but now she had seen everyone. Nothing had been resolved at home or work, and so she had things to do; in fact it was possible her work situation might have deteriorated a bit. She asked him if she could follow up in his clinic; it would be easy to call to make the appointment when she got home.

"Okay, let's talk about that," he said. "Would you be willing to stay in the hospital while we figure this out? And if not, what would you do when discharged, if we could make that happen?"

Winnie didn't have to think about it—that was easy, she wouldn't cause trouble at work anymore, that had clearly been a mistake. She would go home, resume her vacation, finish taking down the east-facing wall, and start ripping out the ceiling too—she was on the top floor, so it was safe, no risk to anyone. "I'm not staying here," she told him, "there is too much to do. I'll just go home and finish my Faraday cage."

He nodded at that phrase, and Winnie asked him if he knew the principle of Faraday cages, that they were conductive enclosures to cancel out electromagnetic fields. He nodded again. "Yes, I use them in the lab all the time," he said. "We put basically mesh cubes around our rigs. The rig is what we build to measure electrical signals in neurons. It's a Faraday cage just like you are building. It blocks noise from other electrical sources that might be in the room, or behind the wall"—he gestured where her gurney had been before she moved it, at the edge of the tiny room—"so we can detect currents, even from single brain cells, even in living animals."

Though still wary, Winnie couldn't help but get a little excited at this connection. She wondered if he knew of the experimental discovery of this shielding principle by Benjamin Franklin, and then the beautiful theorem that had emerged from the physics of electromagnetism, that external fields cannot access the area within a conducting enclosure. That the field creates a compensatory distribution of charges on the conductor, one that exactly cancels out the field itself. A field by its own nature creating its own annihilation. A thesis truly creating its antithesis. "Information suicide," she said.

He seemed to become restless at that, shifting his position in the chair. "So there are some things we're worried about," he said. "You've told me, and everyone, you don't want to hurt yourself—or anyone else actually—and I believe you. But you're destroying your home, and your plan is to keep doing it, because of the worry about your neighbor, that he's downloading your thoughts through his satellite dish. So you're actively tearing your house apart . . ."

Winnie could tell what was coming: they were going to trap her here. She searched his lips as he spoke, looking for signs he was under their control too. Actively destroying her home? This was not true, not at all. She was doing the only thing possible to *save* it.

"I have some paperwork for you—here, this indicates you're going to be admitted, brought in to the hospital tonight, on what we call a legal hold, which we can do, which we *have* to do, for grave disability," he said. "We need to do this because you have a psychiatric symptom that's causing real problems for you, that we call psychosis, which means a break with reality. You're hearing a voice in your head, and you have fears that are not physically realistic, that are causing you to damage your home, and put your own safety at risk."

She felt the world narrowing, going gray except for a narrow tunnel of distorted light around his face.

"It's now our duty to try to figure out what is causing this," he said. "There are lots of different possible causes—and hopefully we can try a medication that could help you." Words came to her mind unbidden, and she tried to match them to his lip movements. *Soapsuds, waitressless, matilda.*

The doctor kept talking for a while longer, then stood up, and she focused back on the meaning of his sounds. He said he would see her tomorrow since he was also working in the locked unit during the day that week, and he left her with a sheet of paper with many words and numbers. *Grave disability*, she saw, and 5150. There was that code from the ambulance. Grave. They had her now. She kept her face still as a fossil, staring straight ahead at the scuffed yellow wall, not daring to picture what lay beyond.

•

The staff administered to her a new medicine that first night, and gave her an information sheet with it, which she kept to study; it was called an *atypical antipsychotic*, and they asked her to sign something about it. Whatever else it did, or didn't do, the tiny white pill certainly knocked her out, and she slept for fourteen hours.

When she awoke, Winnie found herself upstairs, in what they called the locked unit, among a group of fellow travelers, each a refugee from a different sort of storm, washed up onto the same shore. Winnie just listened that morning, not speaking but able to learn from them; it helped that her own storm had made a kind of landfall itself and expended some of its energy already, even by the first morning. She could still hear the *disconnect* voice, but it was less intrusive, no longer a shout—and she was able to focus on people more stably, and follow conversations.

She learned how to slash her arms with a toothpaste tube—she didn't do it, she didn't want to, but she learned it anyway. Two patients were talking in the breakfast area who had done this before—for different reasons—and were comparing strategies like recipes. One, a young woman named Norah, seemed to just want to cut herself a little bit, just to feel pain and see blood, to leave a mark and have it known. The other, Claudia, a large woman who could have been the mother of the young-adult brood, was focused on actual suicide—cutting arteries, letting all the blood out. Claudia was about to start electroconvulsive therapy treatment for severe depression—the doctors thought it would help, but Claudia had a different plan. She was utterly invested in ending her life. All her feelings and thoughts led there, currents joining into one flow that could not be slowed or diverted by wall or lock.

But the unit staff were a step ahead, it seemed—not even a toothpaste tube was available. The nurses were mostly miraculous—with only words and gestures, they managed to maintain peace among twenty altered and demonstrative men and women. The unit was like nothing else Winnie had experienced—a contradictory place, both hard and soft, desperate and secure. And the other patients—she could spend an eternity contemplating their individually damaged worlds. The unit was a maelstrom of fascinating and frightening alternate realities.

Winnie thought about the toothpaste, how the bottom of the tube worked for this job. Its stiffness sufficed; it had the right material properties for sharpenability. She pictured Norah and Claudia each as they had been in other inpatient settings, on less restricted units in the hospital, surreptitiously grinding the end of their tubes on any gritty surface, getting in a few or a few hundred strokes here or there when they could isolate themselves from staff. Winnie thought about how compelling repetitive action could be—with needle or knife—repeating the same action again and again, hundreds, thousands of times. She had an odd idea—that rewarding the act of repetition was the first achievement of the human brain. With relentless rhythm, to make a hard thing sharp: a stick, or a flint, or a bone. Striking again and again, grinding against rock, all through the winter—but with a different goal: to survive back then, not to die.

Winnie picked up psychiatry knowledge too—not from the other patients, but from brief conversations with the psychiatrist who had admitted her—about what they called psychosis. He saw her twice a day, once around eight in the morning in the room she shared with Norah, and then sometime in the afternoon, usually in the hallway when they happened to cross paths. Winnie noticed he seemed as sleepy during the day as he had at midnight. She liked that he liked Faraday cages, and she called him Dr. D. As her storm cleared more and more each day, she began to ask questions.

"Psychosis, what exactly is that?" she asked. "I mean, I think I know, but it's strange to hear you say it—it's an old-sounding word."

"Just a break with reality," Dr. D. said. "It can be used for hallucinations like that *disconnect* voice you hear. It also applies to having delusions—that's the word we use for beliefs that are false, but fixed."

She considered that. "What do you mean, fixed?"

"This fixity part is important," he said. "Delusions can't be reasoned away. Evidence does not help. I used to try, for my patients, when I was still learning. Maybe every psychiatrist has tried—but not for long. The delusion can't be budged. Some patients have these extremely unlikely ideas in impenetrable armor, so they can't be touched."

This idea of the fixed belief clicked with Winnie's engineering expertise. It was like the Kalman filter, an algorithm for modeling complex unknown systems—in which every guess at the value of a system property comes tagged with an estimate of the guesser's level of confidence. And more weight is given, when modeling the system, to guesses with higher certainty. It made sense to Winnie that the brain should work that way too, that knowledge should exist only with certainty tags, and that some types of knowledge of the world—not just the delusional ones—should be trusted all the way to the point of fixity, and placed in the brain within a special bucket called Truth, not subject to hedging or discounting. The category of Truth would allow fast and simple determination of action without clock cycles wasted on statistical computation, and allow the brain to build complex edifices of logic on top of these unquestioned facts. But she didn't say all that to him.

"I think it's not just psychosis that gets fixed like that," she said hesitantly, feeling pressure to get everything from her mind out before he walked away, "but also maybe other ideas." She tugged her Raiders hat down tight—force of habit, really, she was feeling lately that she did not need to wear it all the time. "Like trusting your family, and marriage, and religion, and some kinds of social and political beliefs. It's normal. Every bit of knowledge should have a confidence number attached to it, and some ideas should have a perfect score."

"I guess so," he said. "I think you're right, we do need those . . . rankings, I guess. Confidence estimates." There was an awkward silence. He look down at his patient list, which she knew meant he would soon move on to the undergraduate student, the next room down—blond and smiling and manic and so many words—and never get back to her.

But then he continued. "I think, though, for most ideas about how the world works, a perfect score would not be helpful. And some possible explanations for things are so unrealistic— they should never get close to becoming such trusted facts." He paused again. They were standing in the hallway near the nurses' station, an odd pair, she could see that. She in her hospital gown and Raiders cap, he in his daytime getup of button-down shirt and slacks—one prisoner, one free, and patients meandering around them. And yet a connection was there; they were passing information back and forth, untouched through the noise, on their own local area network. "These unlikely ideas," he said, "should never even get access to our minds in the first place, should never be let loose to rise up into our working active consciousness at all. Do you think you had any ideas like that right before you came to the hospital? Distractions—really unlikely, that should have been just filtered out—before ever rising to the surface."

He was talking about filters, but not quite correctly. In the quieting of her storm, Winnie thought he might be referring to something she had told him in the ER—her story of the screw in the parking lot. She saw now that the idea she had at the time—that the screw had been placed there by Erin to torment her—was quite improbable.

But so what? she thought. Fixity was seen in delusions, but probably was also essential for healthy committed behavior—and similarly, allowing consideration of unlikely ideas seemed to Winnie to be normal and necessary as well. "You know, allowing awareness of something that's unlikely is not a disease," she said. "If you're talking about a filter, you should understand how they work. Optimal filters will still block a few things that you actually wanted to go through—and also will still allow some things that you wanted blocked to instead go through. That's for an optimal filter."

And for ten minutes she described for him Chebyshev and Butterworth electronic filters, and explained how Chebyshev type I filters do successfully block from going through what is not wanted, but unfortunately also block a bit of what is wanted, what should have been passed through. Fine for some electronics, or maybe some nervous systems, but not for the human brain. A species like ours, with survival so clearly based on intelligence and information, should not run the risk of blocking and throwing away potentially valuable ideas.

Other designs, like Butterworth filters, have the opposite weakness: these discard nothing of potential value, but allow too much to slip through. "I think the Butterworth design makes more sense for a human brain," said Winnie, "or, for all the brains of our species considered together. Unlikely beliefs held by some are a sign the species overall is working well." She said she would send him "On the Theory of Filter Amplifiers," Butterworth's 1930 paper. Winnie felt it was actually quite important for him to know that every system operates with an error rate that it accepts, to balance against some other consideration.

"Same with our electrophysiological signals in neuroscience," he said, seeming to agree. "We record tiny currents, and so we have to filter out noise to see the currents, and even the best-designed filters will still block or distort some useful things and allow through some useless things." Winnie had more to say, but at least with that she could let him move on. Now he seemed to know that distortion does not mean disease.

•

The inner voice grew still more quiet over the next day. She also felt decently stable without the Raiders hat and stopped wearing it. Winnie could sense something was getting better, though she felt a bit wary about revealing this to the doctor. He might assign credit to the pill, and conclude that this illness model he had for her was correct.

Dr. D. dropped the 5150 before it expired; Winnie had agreed to stay voluntarily on the locked unit until discharge, since the voluntary unit, the open floor, was full. But she was happy to work with the current clinical team, while the tests continued. She was on vacation anyway, she was learning a lot, and home still didn't feel quite safe.

"There are different reasons people can experience psychosis," Dr. D. said in the hallway, later in the afternoon after dropping the 5150, "and not all have been ruled out yet for you."

"But I thought you agreed," Winnie said, "that there might not even be a problem, it might just be my design. Our design."

"Yes, well," he said, "as you pointed out, people could be designed with different filters, just like everyone has different settings on their sound system. But there's a problem with that idea. . . . This experience has never happened to you before. As far as I can tell, you've always been logical and systematic, with a selective filter—it's maybe one of your greatest strengths, actually. So this whole thing is not really your design."

"What could make things change then, if they did?" Winnie pressed.

"Drugs could do it, but there were no traces of drugs in your system that we found," he said. "Infection or autoimmune disease also, but we found no hint of those in your blood work either. Severe depression or mania could do it as well, but you have no symptoms of these. Schizophrenia, though, has not been ruled out."

Winnie had some sense of what schizophrenia was, and it didn't fit with what she was experiencing. "Doesn't that start in teenagers?" she asked. "I would have had symptoms long before now."

"That's true for men, but twenty-nine is not atypical for first break in women," he said. "First break—that's what we say when schizophrenia declares itself, with visible symptoms like delusions and hallucinations. And sometimes one's own actions can seem foreign, controlled from outside the body—"

"Are there theories for what causes hallucinations?" she asked. "What could be the biology of something like that?"

"Scientifically, nobody really knows," he said. "Some people think that inner voices—like that one you hear—might be caused by one part of your brain not knowing what another part is doing, the brain not recognizing its own inner thoughts as itself. And so your own internal narrative, like the word *disconnect*, gets heard, and interpreted, as the voice of someone else.

"Similarly, you could feel that your actions are not your own, but reflect control from outside. It could just be that in schizophrenia, one part of your brain has no idea what another part wants or is trying to implement, and so an action of the body gets interpreted as a sign of meddling from the outside. The brain—casting around for explanations, which it always does—finds only unlikely ideas, like control by radio transmissions or satellites."

"Wait," Winnie objected. "Why are these explanations always so technological, always beamed information like that?" She had to get to a resolution, and knew she was running out of time again. "You know, why satellites? Doesn't that mean this really isn't a disease? It's more of a recent development, right? A reaction to technology."

"Well," he said, "this feeling of external control and long-range projection of information, of forces acting at a distance, was always a symptom as far as we know, long before satellites, or radios, or any kind of energy wave was known to exist." He started to drift down toward the next room along the hall, in the pattern she knew well now, edging toward continuing his rounds. "I have to keep going now, but I think I can show you how we know this tomorrow."

The next day, as she waited for morning rounds, Winnie wondered if among all the failure modes of the human mind, schizophrenia might be the least understood. She herself had heard nothing explanatory, and felt so ignorant about it, with many gaps and maybe misconceptions. Disorders like depression and anxiety seemed so much easier to map onto regular human experience.

Still, altered reality could also be universal in some sense. In college she had learned that while falling asleep most people can experience brief, bizarre states of confusion and hallucination; she knew that state herself, and that it was frightening enough for the instant it lasts—yet what would life be like if that state came one night and never went away? If that altered reality, once experienced, became fixed? Entrenched and unshakable for days, or for years. The idea was horrifying, and so she stopped thinking about it.

The fragmentation of the self as a concept intrigued her, and was more pleasing somehow to consider—the idea that one part of her could fail to know what another part was doing. The idea made her wonder how integration of the self is ever achieved in the first place. She had always taken this kind of thing, her wholeness, for granted, but apparently it was not so certain. Again thinking about sleep helped her understand, since on awakening, she had always felt an unraveled moment with no reality or self at first, but then experienced a gradual reconstruction, a reweaving. Short local threads—of place, purpose, people, things that matter, schedule, current attributes—came to interlock with long-range threads of identity, trajectory, self. Where is the information coming from, and where is it going, that reweaves the self in those minutes? If that process is interrupted, the result would be an incompletely formed self—and one's own actions would seem unconnected and alien.

As Winnie thought about that disconnected state, a disturbing thought occurred. What if that underlying formlessness—needs unraveled from self, action estranged from plan—is what is real? What looks like confusion and disorganization in those psychotic states, she thought, might be simply the reality that our borders are arbitrary, and our sense of unique self actually artificial—serving some purpose, but not real in any sense. The unitary self is the illusion.

And then, what about that voice, almost imperceptible now? The doctor had implied she was thinking *disconnect*, and not recognizing it as her own thought—but he was missing the deeper point. Even if the *disconnect* thought was "hers" in some sense, who told her to think it? Did she decide, at a moment in time: I plan to think "disconnect"? No, not for that or any other thought. The thought comes. For all people, all thoughts just come.

Only people with psychosis are rightly perturbed by this, Winnie realized, since only they see the situation for what it is. Only they are sufficiently awake to perceive the underlying truth—the reality that all of our actions, feelings, and thoughts come without conscious volition. We all lie on the hard hospital bed prepared for us by evolution, but only they have kicked off the thin blanket, the comforter provided by our cortex—the idea that we do what we want to do, or think what we want to think. The rest of humanity proceeds through life in dumb slumber, serving and preserving the practical fiction of agency.

The next morning, by the time Dr. D. got to her on his rounds, Winnie felt convinced that hers was a state of insight rather than illness. She was not shielded but rather had emerged, and could sense the field, the charge surrounding everything. But before she could tell him, it turned out he had brought something for her, a picture he had printed out—first drawn, he said, by a nineteenth-century Englishman named James Tilly Matthews, in the heat of the Industrial Revolution, in the grip of what they then called "madness." Matthews had imagined something he called an "Air Loom," and drew pictures of himself as a helpless, cowering figure controlled by strings projecting through space from a giant and menacing industrial weaving device. Controlled from afar, by long-range threads.

Winnie was fascinated. So unexplained symptoms and feelings in schizophrenia were just imputed by patients to their time's most powerful known phenomenon for action at a distance—whatever it happened to be that might serve as an explanation—satellite, loom, angel, demon.

Winnie had much to say after that, and she found herself more interested in exploring these ideas than in pressing for discharge from the hospital. Even if she had schizophrenia or something similar, it seemed clear to her that this was not truly a disease, but a representation of something essential—a spark of insight and creativity, an engine driving the progress of humanity.

So the next day, she asked Dr. D. to admit that this could be true, that tolerance of the unlikely and bizarre could be useful—in the context of the human brain and human hand. Only in this way could unlikely things—semi-magical possibilities, concepts unrelated to anything that had ever existed—become real. Such a setup would only be of value to humanity; there would be no value for a mouse or a porpoise in magical thinking, admitting to unlikely possibilities, believing for no good reason that something strange might be true, that a different world might be possible—with no big brain to plan it, or nimble hand to make it.

He was not as excited as she thought he might be. "People have thought about this," he said. "Not to say that's not an interesting idea, or that it doesn't have a certain appeal. It might even be right in some sense. But schizophrenia is much more, and much worse, than a little bit of magical thinking. There are also the negative symptoms of schizophrenia, which prevent patients from even accessing the basic and useful parts of their mental world anymore. There's an apathy, a loss of motivation, a lack of social interest.

"And then there is a symptom called *thought disorder*, in which your whole internal process can become disrupted in a very harmful way," he said. "Think about thinking for a bit, which you have been, but now about the flow of thought. We do plan to think a thing—not always, but sometimes, or at least we can if we wish. We set out to reason through things, we choose to build a series of thoughts: imagining paths radiating from a decision point, planning to go through each of them systematically, and stepping through that sequence. This is a beautiful thing about the human mind, but this beauty can be corrupted. Patients lose the memory of their positioning along each planned path of thought, and even lose the ability to chart the path at all. Words and ideas get jumbled up together, getting inserted or deleted too. Eventually thinking itself is shut down completely. We call that thought blocking—when patients crash out of conversations midsentence, midword. Thoughts come unwanted, but also don't come when wanted . . . and can't be summoned."

Winnie knew she had exhibited long silences in the ER—but she had been thinking about AJ dying. She reminded the doctor about AJ, saying, "I don't think my silences that first day were thought blocking, Dr. D. It was just a strong feeling, from a personal memory that mattered—my brother's death everyone was asking me about, nothing else."

"Okay, yes, that may not have been thought blocking," he said. "It looked that way—but the good news is it's happening a lot less for you on the antipsychotic medication. And thank you for letting me know. We try to visualize what is going on inside our patients' minds—but thought disorder is not something most people can vividly imagine, and so we could get it wrong. It's maybe even the most debilitating symptom in schizophrenia, but extremely hard to explain."

Maybe because this is the most human symptom, she thought, a deficit in the most advanced brain system, with no analogy in any other animal or being. But more importantly, control over one's own thinking is just an illusion anyway—it's the fantasy of control that is uniquely human. Thoughts are only ordered after our guts decide what we want, and fictional thought sequences are built up and installed retroactively. This perception of order in our thinking is as unreal as agency over our actions. Both are rationalizations—just neural backfill.

•

The day before discharge, he came to update her on the final reading of her MRI. There was nothing in her brain that they could see—no AVM like the one that had killed her brother, no tumor, no inflammation. "What this means," he said, "is that your episode of psychosis might well be a sign of schizophrenia. We don't know for certain yet, but that is the working diagnosis. But there's one more test we need to do. We need to check your cerebrospinal fluid for signs of something that might be treatable—cells that shouldn't be there, or infectious agents, or proteins like antibodies. This means we have to do a lumbar puncture—a spinal tap."

Winnie felt herself flinch slightly, remembering the terrifying length of the chemo needle. "I know, sorry," he said. "You've had these done before—yes, it's invasive, but almost painless, and we know from the brain imaging that you don't have any worrying pressures in there that would make it unsafe." Her experience as a teenager fully surfaced itself, uninvited, as he prepared the consent form. Winnie remembered how she had been positioned on a bed facing the wall, in a fetal curl to present her lower back—but it was true, she remembered no pain, just a deep and achy pressure.

"It is pretty unusual to do on this unit, so we'll have to take you to the open floor," he said. "No needles are allowed on the locked unit, except in emergencies." Winnie signed the consent, they had her change into a hospital gown, and then she walked with Dr. D. and the nurse to the locked exit door. The ward clerk buzzed them through, and she was out in the legal open for the first time since her admission nearly a week ago.

As they set her up in a procedure room, she considered the irony of what was about to happen: after her frenzy of concern about long-range access to her brain, here she was willingly allowing them direct entry, right into her central nervous system. And they would withdraw material—her own liquid from deep within her—and keep it, and test it, and enter the results into databases that would never go away.

But she had somehow consented, and it was all happening. Dr. D. positioned Winnie on her side with a gentle curl, and with her hospital gown pulled away to expose her lower back. First came the surface anesthetic: a small stick, from a tiny needle. The big one would come once he had the location exactly mapped with his hands. He talked her through it—"I'm finding the boundaries . . . framing the top and bottom lumbar vertebrae, these define the space, the fourth, the fifth—there it is." After a breathless pause, she felt that familiar deep ache. The needle was in her spinal column.

It would be a clear liquid, she recalled as she fixated on the wall in front of her—cerebrospinal fluid, unlike any other in the body. They would test it for cells, sugar, and ions. CSF, bathing the brain and spinal cord, cushioning the neurons of thought and love and fear and need, with just the right salt concentrations of our fish ancestors, along with a touch of glucose—a little bit of the ancient ocean we carry with us, sweetened, always.

The next morning, he conveyed the results: more good news. Nothing of concern, all clean; in fact, he confided, it had been a champagne tap—which meant the CSF had come out fully clear with no blood from a nicked capillary, not a single red blood cell. For residents and interns performing their first LPs, he said, this is usually occasion enough for a bottle of champagne, marking a milestone of technical skill along with a little luck. But more important for Winnie: no white blood cells, no inflammation, no proteins, no antibodies. Glucose and ions all normal.

Another minor side note: something called *cytology* was still pending, a detailed analysis for cancer cells, but lymphoma recurrence was not suspected by the laboratory. And so this day would become the day of her discharge, as he had promised—and they would send her home with a prescription for the new medication, the antipsychotic.

"And the discharge diagnosis?" she asked. "Will you say schizophrenia, or not?"

"We still can't be certain, but schizophrenia is likely," he said. "Some psychiatric diagnoses can be applied only if everything else is ruled out, only if enough time passes with no other explanations found. So for now, we'll give our temporary diagnosis: schizophreniform disorder, which can be converted to schizophrenia at your outpatient follow-up." An unappealing prospect—Winnie felt disinclined to let that happen.

Champagne tap—my brain feels like champagne, she thought later, back in her room, waiting for the discharge orders to go through. She had liked that phrase he used, *champagne tap*, and so she began to play with a more retro image of filtering—moving away from modern electronics, to more of an Industrial Revolution filtering of bubbles, more like James Tilly Matthews might have pictured as he pondered his drink. Bubbles of ideas are seeded deep, guesses to explain the world—

why is that screw there?—models nucleating on the side of the mind's champagne flute, rising quickly if able to combine with others in support to form a larger bubble, a more complete hypothesis, that can rise more powerfully past filters that can only arrest the small and weakly moving, the unlikely, the poorly justified.

The bubbles that rise the fastest and grow the biggest, encounter more support and reach the brim—the border of awareness—only then to burst into consciousness. Once that burst happens, it's irreversible. It's no longer a guess, it's Truth—molecules forming part of the oxygen of the mind now. There is no re-forming of bubbles that is possible; there is no sending them back into the champagne.

And most important of all—sometimes a few little bubbles that should have been stuck instead slip through. Winnie thought: Why not send them up? The world is always changing.

She was discharged on the afternoon of her tenth day—the last dose of the pill, the antipsychotic she had been given daily from her initial admission, had been administered by her nurse the night before, and she had a prescription to fill at home, so she could keep taking it. With a tentative diagnosis—schizophreniform disorder—she was free to go.

•

Winnie never filled the prescription or followed up in the clinic, and never planned to. She felt fine. When she got home she skimmed Dr. D.'s card across the room and let it lie where it fell by the gas fireplace, a white marker thrown down where she could see it and remember—and in the meantime there was work to do.

She felt good going online, not even worrying about Erin. The hacking conspiracy was still there in her mind, but not as an overwhelming invasion anymore—more of a polite houseguest. They could leave each other alone, pass in the narrow hallways of her mind with a slight turn of the shoulders and a courteous nod.

She even felt more secure about her own body, her own borders. The Raiders hat went back into storage. As she was reorganizing the closet she came across her old copy of Benjamin Franklin's "Letters and Papers on Electricity" from 1755 and went straight to her favorite passage, from his letter to Dr. L——, describing the discovery of what would become known as the Faraday cage, savoring again, as she read his words, Franklin's false humility:

I electrified a silver pint can, on an electric stand, and then lowered into it a cork ball, of about an inch diameter, hanging by a silk string, till the cork touched the bottom of the can. The cork was not attracted to the inside of the can as it would have been to the outside, and though it touched the bottom, yet, when drawn out, it was not found to be electrified by that touch, as it would have been by touching the outside. The fact is singular. You require the reason, I do not know it. Perhaps you may discover it, and then you will be so good as to communicate it to me.

Winnie felt a connection again to the cork. After a brief and tumultuous emergence, where she had been buffeted by the fields of an external reality, she had now returned to the silver can, the shielded cage, the shared and common human frame.

There probably never was a miscarriage though—that idea had become uncoupled from her also, drifting off, a cinder lost, a dark mote dimming.

She ate ravenously that first week home—with a hunger like nothing she had felt before. Controlling her own food again was a revelation, a release. She cooked pasta, bought cakes. Toward the end of that first week, an odd thought appeared—she was not sure she had a mouth. Even while eating—especially while eating—she had to touch her lips to make sure they were hers, and were still there.

Between meals, the patent lawyer in her reemerged—strong and refreshed and tireless. Just as at work when tackling a new field of art, she spent many hours each day at her computer, delving into the scientific literature, seeking knowledge and precedent. She found her way to dense and intriguing papers on schizophrenia genetics: the collection of DNA sequence information from human genomes, with massive teams of scientists spelling out individual letters of genetic instructions within tens of thousands of schizophrenia patients. She wandered, fascinated, through the hundreds of genes found, associated, linked—that all seemed to play some role in schizophrenia. Each gene alone had only a tiny effect on the individual human being, with no single thread setting the pattern, none by itself defining the weave, or the fray, of the mind.

Instead, all the threads together manifested health or disease: only in unity did they form the full tapestry. It seemed to Winnie that mental illnesses—schizophrenia but others too, like depression, autism, and eating disorders—even though heavily genetically determined, were mostly not handed down across generations like a watch or a ring, nor like the single genes controlling sickle cell or cystic fibrosis. Instead, in psychiatry it was as if the risk were projected forward as a set of many vulnerabilities from both parents. Each person's mind was created by thousands of crossing threads, intersecting orthogonally and forming patterns diagonally, to create the twill of the individual. There were genes for proteins creating electrical currents in cells, genes for molecules at the synapses controlling information flow between cells, genes for guiding the structure of DNA in neurons that would direct production of all the electrical and chemical proteins, and genes for guiding the long-range threads within the brain itself, the axons that connected one part of the organ to another, on an inner loom of interwoven wiring controlling everything, directing all aspects of the mind, setting traits and dispositions like tolerance for the unlikely and the weird.

In some people, Winnie realized, when the warp and weft intertwine just so, a new way of being is allowed—a pattern coalesces with just the right or wrong set of threads. Hints of what might come can be found on both sides, forming the family tartan, in those predisposed. Looking back, partial motifs can be discerned among the vertical or horizontal threads—human traits as proto-patterns. In both lineages, there may be found uncles or grandmothers who were just odd enough, who could let their minds relax the vise of illusion, who could loosen the grip of an old paradigm, to close, firm, around a new one.

And the stronger the old paradigm—with more societal inertia—the more certain these outlier human beings had to be of their new outlook. Their convictions had to be fixed—once shifted, never letting go—committed for no good reason, since there was none. For who can defend the new and unproven against the old and established? Only the unjustly certain—who believe to a level never provable, who already must dwell a bit apart and aside, who already can access now and then the fixity of delusion.

But when two highly vulnerable lineages converge, a person might emerge who is too disengaged, allowing too much through, having lost control of thought—or rather lost the reassuring illusion, the perception of thought's order and flow. A shaken human being is formed who cannot decide which paradigms to abandon, or which to never let go—who cannot even pretend to decide anything, any longer, amid the stirred-up turbulence, the swarms of bubbles effusing and bursting forth unchecked from the champagne. Then all the bubbles become spent, and the human being ends with the negative symptoms Dr. D. had described—avolitional and flat.

As Winnie read more about severe schizophrenia, she found it harder to preserve that idea she had as an inpatient, that there could be some benefit to the disease—for those suffering, or for their loved ones. It seemed that the most insidious symptom Dr. D. had described, the thought disorder, inexorably progresses if untreated, until utter disintegration. Thinking becomes more and more distorted, until the mind cannot keep track of obligations and connections, and loses emotional range, both highs and lows. Gone is any urge to work, to clean, to connect with friends and family. The mind becomes awash in chaos and terror, the body frozen and catatonic. If left untreated, the patient's life ends in confused and bizarre isolation, with the duration of any planned thought shrunk down to a few seconds or less before its annihilation.

Winnie remembered vividly something the doctor had said in the hallway, in their final conversation, when she had been repeating that error need not mean disease. "A group in which some people tolerate unlikeliness this way may do well over time," he had said, "but don't forget—some will suffer terribly." Now in her apartment she wanted to respond, but it was too late. She wanted to tell him that she understood now, and that this was not just true, and important, but should be taught to the community—to advance understanding, to elicit gratitude even—so that all could truly see the people who were ill, to understand their burden borne for us.

He would probably agree, but what he would not like was another thing she wanted to say, of which she was just as certain—that we all need delusion as individuals now and then. She wanted to tell him that within each person there should be a breakdown in reality at times. We should recognize this need, in ourselves and each other, and move with it like music, and sweep each other along, leading or following as life suggests, since there is not one reality that works for every decision in each phase of life, for every pair or group or nation. We have brains and hands; we might make our delusions real.

And she imagined his rejoinder already, since like any good lawyer she could play his side just as well: that this was fine and romantic to imagine, but one cannot make anything real, create any complexity, without controlled thinking, the ability to plan many steps—and schizophrenia shuts all that down. Evolution has not worked out how to consistently protect everyone from thought disorder—leaving human minds with a vulnerability especially destructive in the modern world. Simple and small primate groups may not have needed thoughts to flow in sequence for long periods of time—but the stability of our community structure requires people to live and work together over long timescales, and allows multistep planning to matter.

Winnie knew this perspective had to be at least a little bit right; she had found plenty of data to support the idea that civilization contributes to the problems caused by schizophrenia—including evidence that disease symptoms are more common and strong in city dwellers. People with only mild genetic predispositions could still, it seems, be pushed over the edge into psychosis by other risks and stressors of modern life. Winnie also found many accounts of perfectly healthy people who became psychotic only after their first cannabis exposure—and of others with seemingly pure mood disorders like depression, who experienced delusions only because of the mood disorder, not schizophrenia. She thought these human beings probably all had at least the proto-pattern, half-woven. With a tweak from the environment, a toxic chemical, a stress from city living or social disruption, an infection—

whatever it might be, Winnie thought, a second hit on top of genetics can complete the pattern, and change reality.

Two hits: this was a concept she was familiar with from cancer. Winnie remembered asking her oncologist as a teenager, why her? Why not Nelson or AJ? Why not her best friend Doris, who smoked secretly every chance she got? Maybe the two-hits hypothesis could explain this, her doctor said; maybe Winnie had some vulnerability from genetics, but mammals have two copies of every gene, and other kinds of backup systems too, so some second hit was needed to allow the cancer to happen, through another change in her DNA. It could have been a cosmic ray, a long-range particle traveling from the sun, or a gamma ray from another galaxy even, traveling through space for billions of years, and hitting one chemical bond in one gene in one cell of one young girl in Wisconsin. This was happening all the time to everyone, but in Winnie's cell there was already another problem, an altered gene from birth. A disruption came atop another; it was a double tap, too much, and the system tipped over into the uncontrolled growth of cancer.

Nobody knew if the two-hits idea was right for mental illness, but Winnie thought it could be. The science was not there yet in psychiatry, that much was clear as she spent nights reading the papers and reviews. The biological knowledge was limiting in this field, though there were a few insights. There was altered communication across the brain in schizophrenia, shown with methods for imaging brain activity in people. Parts of the brain were not keeping the other parts updated. There even had been observed, during hallucinations, an altered synchrony of activity across the organ, like one hand not knowing what the other hand was doing.

Winnie had so many questions, so much to say, and nobody to listen. She remembered he might have said a patient's break with reality had brought him to psychiatry in the first place. Not that it mattered, but it mattered, and she wanted him to know that it did. We take our shared reality for granted, and our reaction to that illusion, and if she could ask him to do something, it would be to let the world know a simple Truth: our shared reality is not real; it is only shared.

•

In her second week home a goal emerged, a god took form, a mango ramjet. She would write to him, in a detailed letter, by hand, in unerasable black marker, in all caps so nothing would be missed, with everything she never had time to say, that she hadn't known how to say clearly.

She would tell him more thoughts, mote thoughts. There was a dispersed element, moonlit underdrumming, a nocturne. Java Pajama Princess was her new name, that was something to tell him. He might not understand, he was unbearded, an unjesus. He would write back with his full name, not what the nurses called him on the floor, that false note, popish. No, his full name, and she would tell him so, she said she was not of Dravidian ancestry and did not appreciate the implication—misogamy—her voice cracking, turning to a weak whisper even as her helpless anger grew—what was he implying. Not one kilogauss of influence upon her, she was pure and free, not some ropedancestepdancing firebrat. Was she eating too much? Lickerish. She was double-tapped. The influence was coming, the outlet was not easy or east but westnorthwestwardly. She paused, took a breath, and apologized. A torsade. It was none of her business what he was trying to imply.

Her phone went off; something clenched her deep. Fillet the firstborn, the fistborn. It was him. Winnie reached out to the phone, but hesitated. The other side of the screen. She let him buzz through to voicemail. An hour later she played the message on speakerphone, after she felt the phone's capacitors had fully discharged. The cytology report had come back, from the spinal tap, that last formality: "*Rare highly atypical lymphoid cells, consistent with prior material, involvement by a T-cell lymphoma.*"

Her engine of a brain revealed at last its dark secret. Covered over but always there, her frailty had been lying in wait, like AJ's AVM. And then came the second hit: for him, the pressure surge; for Winnie, it was cancer cells, stirring up the champagne bubbles, swimming in her vulnerable sweet sea.

She settled to the floor, reaching out again to AJ's last day. It was not hard; the air loom projected across time as well as space. And she knew the threads that mattered; some of them were hers. *When he saw the bank clock, AJ knew he would have to jog the rest of the way. As he ran he looked down at himself and his shirt. There was some dough baked on there and he tried to brush it with his hand—most of it came off but there was still some white stuff that he couldn't wipe away, and his hand was sweating, and it all made things a little worse. He should have brought another shirt. He kept up a steady pace, trying not to exert himself too much as he neared the bank, and jogged across the South Main intersection and into the plaza, skirted the fountain, and ran through the glass doors just behind a guy on crutches. He saw the elevator but no time, up five floors two steps at a time; he walked quickly down the hallway, checked behind him to make sure he wasn't leaving flour footprints, and stopped just outside the office waiting to catch his breath. Wiping his forehead he looked around at the walls and the ceiling; the hallway was very clean and brown. He thought about the frozen yogurt girl next to the bakery and her hair, curving up like a cinnamon roll, firm and brown. He thought about how her eyes had circled around his face, like a nervous blue jay, when he had asked for her number. After a minute he reached for the door with a shaky feeling inside, watching the dim and shadowy reflection of his face in the glass panel of the door, feeling he was at the crest of a hill, carrying over the summit a large piece of cardboard in his sweaty hands, to slide it down the summer slope like he did with Winnie and Nelson when they were kids. He was going to see how things looked, on the other half of the world, after a long climb getting ready to coast down. The cries of triumph and pain of the other climbers were fading into nothing for a moment . . . as if in respect for this instant. The door was locked; it took AJ a bit to realize the door was locked. It was strange—the knob turned but the door wouldn't open. AJ trembled and tried again. He stepped back then, trying to think what it meant. His eyes looked for some message or note or clue, but there was nothing. Wrong office maybe. He reached for the appointment card in his pocket, but it was the wrong card, it was the mechanic's. He'd brought no number, going to miss the appointment it had taken him months to get. His head* twinged. AJ pressed his hands to his temples as he walked back down the hall. He took the stairs slowly, knees buckling, feeling a strange and surging flood. The lobby was lost in a black fog. Scared, he walked as steadily as he could through the lobby and out the door. The sun was hot, but dim. His legs and arms were shaking but he made his way slowly to the plaza fountain. He walked around the spray, unsteady, and waited to cross South Main, watching the faces in the cars as they passed him by. He went to his knees. He remembered a bird he had seen collide with a glass bus stop once. It had beat the dusty pavement with its wings for a while, unable to lift off, and then just looked and watched the other birds fly by, intent on their own lives, haloed by the sun, to mate and feed and build and sing. Twilight seemed to be deepening over everything. He thought that he might see the frozen yogurt girl if he could just get back to the bakery. I would like to stay there with her, he thought. It was a slight slope downward; if he could get up all he had to do was swing each foot forward, one after the other, and he could just almost coast. All the faces in the cars going home . . . The door would not open. The door was locked. The pain in his head rose and spread. So clean and shiny the glass was everywhere, it looked as though it wasn't even there, the bird hit and the glass was everywhere. The hallway was long and dim, firm and brown. It was not easy to see seeing again. Sort of a dove, the bird had reminded him of Winnie, he had been so worried for her. As he bent over, the bird looked straight up at him, just like Winnie would, steady and the only one who would. Waiting for it to pass, he clenched his eyes shut, waiting for it too. From his knees he fell down flat, and then she was there with him, smoothing his forehead in a gentling of wings.

CHAPTER 6

CONSUMMATION

Farewel happy Fields
Where Joy for ever dwells: Hail horrours, hail
Infernal world, and thou profoundest Hell
Receive thy new Possessor: One who brings
A mind not to be chang'd by Place or Time.
The mind is its own place, and in it self
Can make a Heav'n of Hell, a Hell of Heav'n.
What matter where, if I be still the same,
And what I should be, all but less then he
Whom Thunder hath made greater? Here at least
We shall be free; th' Almighty hath not built
Here for his envy, will not drive us hence:
Here we may reign secure, and in my choyce
To reign is worth ambition though in Hell:
Better to reign in Hell, then serve in Heav'n.

—From JOHN MILTON, *Paradise Lost*

The medical student and I began to take our leave. Our first ninety minutes with Emily had brought no understanding and revealed no useful role for hospitalization. She had been directly admitted to our open unit by the inpatient psychiatry director, leaving me out of the loop in determining if admission was a good idea.

Emily was eighteen years old, legally an adult but much younger than our other inpatients, and would have been sent to child psychiatry if she had shown up just a few weeks earlier. The initial chief complaint—*unable to sit through class*—was actually her parents' complaint, and to me this situation seemed better suited for the children's hospital than for our acute adult inpatient service.

Over the course of the intake examination, we discovered that Emily had been a star student, but the full fifty minutes of a class period had become too much; at the beginning of the school year she had somehow developed the need to get up and leave class halfway through, and then over a month or so, this had progressed to the point that she couldn't go to class at all. Nobody knew why, and she would not say. We did learn from her that she was well versed in poetry and literature, and had won trophies as a softball pitcher and competitive equestrian.

During our interview, the orthopedic surgery ward clerk had paged me several times about a patient of ours who needed transfer orders to come back to psychiatry after hip surgery. At this point working with orthopedics, as peevish as they were, seemed more productive than continuing with Emily, since they wanted something I could provide. We began navigating around the chairs toward the door of Emily's room—trying not to seem too hurried, promising to return.

"One more thing," said Emily, and I turned back from the door. From her cross-legged perch on the tightly made bed, she was stretching her arms above her head, arching against the sunlight streaming from the window. "I really don't think I should be alone right now."

Oh. Well, here we go. Now the reveal; the inner storm would finally break. I waited, not asking.

Emily's blue-gray side-eye contact with me was accompanied by a quarter smile. She said nothing more. The silence stretched and filled space too. The pressure built, but no cloudburst came.

I looked around the room for insight. There was oddness: her still-packed suitcase, and her laptop and phone neatly stacked on the bed-side table—personal possessions that were not a typical sight even on the open unit. But this I understood—the whole sequence of our usually choreographed intake process was off, due to the unusual nature of the admission. She had just arrived and hadn't even been met by the charge nurse yet.

I looked back to Emily. I had waited for her to continue longer than I normally would, deliberately modeling for the medical student how to let the patient declare herself—demonstrating how to not pre-frame whatever it was into something else, to inadvertently morph the underlying issue into an object of our own making.

And then the silence finally became noise in itself—negative, distracting—even a bit hostile. "All right, Emily," I said. "Let's talk about that." There was no choice but to move back into the room, with my student in tow. We returned to our chairs and sat, white coats settling around us like falling marionette strings.

Not only had our history-taking failed to divine a serious psychiatric condition, but Emily's outpatient lab tests had also been normal—no hyperthyroidism, for example, from Graves' disease, which could have explained agitation and restlessness. With so little information, my diagnostic thoughts felt scattered and poorly formed, mostly relating to anxiety—maybe a social phobia or panic disorder. But she had not endorsed any anxiety-related symptoms. I had also considered ADHD— and had ticked through the symptoms associated with this term, one of many evolving frames we use in psychiatry for states we are still working to understand. As insights come from research, we know our models and nomenclature will be revised and discarded and replaced in a generation, and then again in another. Yet these we use because they are what we have now, helping to guide both treatment and research; each diagnosis comes with a list of symptoms and criteria. Emily was endorsing none of them.

All my direct questions probing these possibilities—and even my less direct methods, like open-ended pauses needing to be filled by the patient—had unearthed nothing substantial. She had some mild depression but never suicidal thoughts; a few hints of the eating disorder traits so common in her age group; and a touch of some obsessive-compulsive qualities. But we hadn't been able to address the core problem, the chief complaint; we couldn't explain why she could no longer stay in class. Only as we were headed out the door, thinking our diagnosis would have to be a placeholder only—*anxiety disorder not otherwise specified*—did it seem the real conversation started.

And now, with her cryptic reopening of the interview, new diagnoses sprang forward eagerly like racehorses charging from the starting gate—but then all stumbled and crashed into one another. The straightforward diagnoses were somehow even less coherent now. If she were intent on suicide, she wouldn't want someone to sit with her. If she were psychotic, she would be less organized, more cagey. And finally, a borderline patient would not be so diffident, and might have led with abandonment more directly.

Whatever disorder lay within was both subtle and strong; she looked physically healthy and did not seem to be suffering, but something had overtaken her powerful mind. At this crucial moment in her development and education, Emily's greatest strength had been taken away; her passport to the future had been lifted from within her, by a light-fingered entity she knew, a thief she was protecting.

As her last words hung in the air between us, something else happened to her, to Emily's scholar-athlete self as shown to me, to her robust and brash façade. For an eyeblink, the mask flickered and fell, and in an instant, all was really real. Though she had spoken a truth as she knew it, there was also a slight twist at the corners of her eyes and mouth. She was showing me something, and it was almost funny . . . but not showing too much, because, well, she was still a teen, and it was still embarrassing.

"Why should you not be alone, Emily?" I asked.

She said nothing more. She was tracing shapes with her finger on the thin and tightly drawn bedspread, watching me out of the corner of her eyes. Emily had spoken something important, and yet it seemed there was also a secret joke unexplained, one that she was tempted to share. Was this all a deeply disguised malingering, from a clever manipulator of the system, working for some gain I had failed to perceive? Or was the humor darker than I could imagine—morbid commentary on a destructive side of her, with desire for self-harm—a cloaked wraith that she had been fighting but could not bring herself to disclose, at least until a social loosening brought on by the moment of our departure.

Ten seconds of silence. What next? I had an ally here, Sonia. I looked over to her.

Sonia was the medical student but also a sub-intern—advanced, and tasked to behave like a full-fledged intern, playacting at the next level as if she had MD authority to make treatment plans and write orders. Sub-I's were expected to perform the doctorly part in each scene right up to the moment of actually signing each order—a challenging role-play, designed for medical students who have decided their specialty, heard their calling, and now are seeking a head start in experience. It's

a difficult line to walk, acting authoritative without true authority—
requiring self-confidence, social smarts, and a tendency to be right.
Strength.

And Sonia was strong—fearless and resourceful, quick with pen and
phone, adept at making things happen. It had been obvious right away,
in her first moments on the team—though I tried not to categorize
people quickly or absolutely, having come through medical school in a
harsher and more binary time, when swift judgments were routinely
made by the team as each new member rotated onto the inpatient care
service: a new face, a blank slate not chosen or known before by anyone
present, but thrust into the midst of urgent life-and-death decisions.
When I had been at her stage, nobody on the team really cared about
how creative the new student might be, or the quality of the papers
published—all of that was irrelevant. A wholly other categorization
came into play that never had existed before in the life of the medical
student. Unforgiving labels were everything: was the new student
strong, or weak?

Teams coalesced on snap judgments, right or wrong but made
quickly. Medical students generally suspected little of the importance
of their first few actions upon joining the team, but in that time they
would earn a label—one way or the other, spoken or not. All was not
lost if things went the wrong way on any one team, since the students
would rotate off the service in a month, moving on to new roles, new
growth, and discovery of new strengths—but that month of time would
remain frozen thus for those who were on the team, never to be un-
done. In low moments I wonder: in how many senior physicians' minds
am I still stored in one of these categories—as strong or weak, and nothing more? Before meeting Emily, when I was a medical student starting my clinical rotations—and front-loading the surgery rotations,
since I was sure my residency would be in neurosurgery—there had
been plenty of opportunity to show weakness.

My head was still in the clouds from my PhD, which had been in
abstract neuroscience and so nonclinical in any sense, and I was more
than a bit defiant—stubborn and unwilling to accept or work with the
axioms and rituals of medicine. In my resistance I was hesitant with

medical custom—and yet sometimes my style, by chance, fit the team's interests. In an early vascular surgery rotation, I had no idea what I was doing, but happened to ask an interesting (if slightly irritating) question on the first morning. As a result later that same day, on afternoon rounds, I was introduced by the chief resident to the attending as "the new medical student, strong." The attending said "Good." How wrong they were, but after that nobody bothered me—I was in, it would be a good month. The student was strong. The team, now set and labeled, swept on.

Later, in my resident and attending years, I thought of myself as part of, and supporter of, a changing culture in which some complexity could be tolerated—in which doctors recognized that the world needed more than one approach to doctoring. Sonia was not weak by any measure though, so when I looked over at her, not knowing what to do, it was for any of her many strengths she could bring to this nameless domain. We had been together for two weeks on the same inpatient team, and we'd had time to get to know each other. She had the same sort of provenance as Emily: similar academic upbringing, diverse and literary, quantitative.

We exchanged a lot of information in that moment—Sonia was keeping quiet, as was I, but her slightly widened eyes, locked to mine, indicated we should explore more deeply.

Looking back at Emily, I picked up no fear, no panic, no anger. Rather, she exuded a kind of nervous excitement, as if she were getting ready to step out on a first date—or no, more like an affair—and then I knew. A representation of sorts, of Emily herself, could be projected onto others I had seen and stored away inside me, from my time long ago on the adolescent psychiatry locked unit—and with only a little warping here and there, the images aligned perfectly.

There was another being in that room with us, one that she needed, feared, and could never leave. Emily opened up and showed me because it didn't matter, there was nothing she nor I nor anyone could do. She did have a ferocious date planned; it was happening and nobody could stop it—but she wanted it known and witnessed. This was a straightedge, unaltered, unsophisticated truth she spoke—one generation stating a hard fact to another—only telling me of the world as it was. The fact was this: she didn't want to be alone, but I should be the one afraid.

•

By that point, I had treated many patients with eating disorders. I'd spent months on the children's hospital locked unit, which is effectively a dedicated anorexia ward, where I had seen patients from mildly ill to near death, and heard the diverse kinds of language the teens used to describe anorexia nervosa and bulimia nervosa. Some patients on the mild side of disease even personalized the two disorders as Ana and Mia, but most patients on the severe end abandoned all pretense of metaphor for their illness.

The psychiatrists working in this realm have deep intelligence and experience, yet their constructions (as with much of psychiatry) are unmoored from the bedrock of scientific understanding, and I had found no greater mystery than eating disorders, anywhere in psychiatry or medicine. None greater in all of biology.

With Emily, I was cautiously aware of a particular priming to consider this kind of diagnosis, since at that same moment I had other such players on the open unit, other patients in the same domain. Micah, for example: art-dealer kibbutznik, eyes dark as shoeblack. He had a sharp and closely trimmed Vandyke, and was frighteningly thin, with a tube snaking up his nose and down his throat. Micah lived in a very deep and severe relationship, with both diseases at once, anorexia and bulimia. Dangerously extreme weight loss resulted, and the contradictions and conflicts were draining. It had become full-time work for Micah to meet the demands of both diseases, to give each the time needed.

Anorexia nervosa is often personalized as cruel and strong, a duchess-like mean girl, distant and stern, locking subjects in a cold tomb of cognitive control. To assert independence from a survival drive, and to reframe the drive to eat as an enemy arising from outside the self, anorexia has to become stronger than anything the patients have known or felt; and the patients start strong themselves—they would have to, in order to manifest such a thing.

With anorexia, they control the progress of growth and life—and so of time itself, it seems. Anorexia prevents sexual maturation in the younger patients, slows aging, and is not cured by medicine; no drug can liberate patients from its grip, thus forcing desperate measures. When we were most acutely worried about Micah, watching as his heart rate and blood pressure dropped to astonishingly low levels, he would allow us to insert a nasogastric tube to pour some calories directly into his stomach. But he would then rip out the tube as soon as he was alone, sometimes before we had a chance to get anything in, so that we had to go through the process of replacing the tube. I could almost hear anorexia mocking me from inside Micah's mind as we went through these motions, as he watched impassively, all three of us knowing what I would do, all three of us knowing what he would do, the two of them secretly smiling, laughing at the tube-wielding drug-mongering chucklehead.

But bulimia nervosa is different. Bulimia brings crazily exciting reward—not suppressing food intake to the minimum but pegging it to the maximum—binge and purge, and binge again. Bulimia seems to create a more positive bond than anorexia; bulimia can scratch an itch deep under the skin, leaving the appearance of purity and health while providing the rawest of rewards. Nothing limits how much bulimia can give you, except how much potassium you have left in your frail and contorted body before you die. In all its forms, bulimia knows what you really want, will excite and hurt you more ways than anorexia, and will kill you just as dead in the end.

Mortal allies and rivals—anorexia and bulimia nervosa—are each hated and embraced, each a snarl of disease, deception, reward. They dwell further from the reach of medicine and science than most psychiatric disorders, in part because a partnership of sorts takes root between patient and disease. Sometimes crush-like, sometimes hostile, sometimes only practical—the partnership with the patient is forged, like many interpersonal pairings in the real world, from a living dialectic of weakness and strength. And though no drug can cure these two diseases, any more than a drug can erase a friend or an enemy, words can reach them as one human being reaches another.

That these disorders are strong, and can be imbued with personhood, creates a situation unlike any other in psychiatry, or in medicine more broadly. Addictive drugs—in the setting of substance-use disorders—come closest to this perception of a controlling external power, though with less personal connection. Eating disorders exert both forms of power: governing authority and personal intimacy.

The power of either anorexia or bulimia nervosa, as with the compulsion of drug addiction, still can derive from an initial, even momentary, consent of the governed. Later this authority becomes malevolent; freedom is lost as time passes, and patient and disease move close and closer—until like any stellar dyad, twin suns spinning around each other, they become locked in a gravitational well, a hole deep and dark, destroying mass with every cycle, collapsing into a singularity.

On the pediatric ward I had seen anorexia nervosa in its most severe and devastating form—a disease dwelling mostly within teenage girls, with both patients and families consumed. These were uniquely deadly dynamics I saw, mixing love and anger, with parents frantic to feed their young, full of fury at this inexplicable monster. Families would blame each other, with hints and digs and clawed swipes and violent detonations, since there was nobody else in reach, and no other way to make sense of their emaciated child, surrounded by yet refusing food. There is no clearer example in psychiatry of human suffering that would be addressed just by understanding—even without a cure.

These were children who had been so strong—stars and performers, disciplined across dimensions, utterly beloved—and yet so starved that their brains themselves were dying away, beginning to atrophy, shrinking and peeling back from the skull inside. Children who had become so fragile and cold that their hearts were slowed to forty, or even thirty, beats per minute, with blood pressure hard to measure, hard even to find—the biology of life slowed and almost frozen, maturation arrested and even reversed, the disease-patient dyad rejecting the impositions and effeminations of the teen years—age, adulthood, weight—those shared enemies, fusing into one and denied as one, rejected as a force from without. Children in mid-teen years with preteen appearance and demeanor—and yet socially smart, even in the depths of the disease still verbally swift, adept, expert navigators of cliques and culture, deft at argument, while failing at that most simple math: the basic topology of survival, the taking in of food.

Many come near death, and some die. Why, ask the families, please tell us.

Why not start by asking the patient, the host of the disease? Anything verbalized would help our understanding, even if (or perhaps especially if) in the uncomplicated language and perspective of a child. But symptoms are hard for patients to explain in anorexia, as in any psychiatric disease. We can no more expect an explanation from the anorexia patient than from a person with schizophrenia when we ask how a hand can feel under alien control, or from someone with borderline when we ask for insight into the exhilaration and release of cutting. Some people simply cannot exist as others wish.

As family and doctors try to step in, to intervene, the patient-disease pair develops deceptions and dodges, whipped hard and harder from within. Together they have reframed desire, reshaped the meaning of need—as can happen with meditation, or with faith—but unsustainably. Anorexia is strong but causes fragility, and defends itself lethally. Anorexia preaches loudly in front of the mirror, and then later, off the pulpit, still whispers relentlessly with sibilant words learned in secret— a mimic, a hustler, a charlatan within—until in the end the lie is accepted. The pretense first gains leverage for its utility, but then grows rapidly to meet the monumental scope of its task. Once commissioned, the neural mercenaries cannot be recalled, but spin out of control into a rogue army ravaging the countryside.

These are not simple delusions. In the end, the patient somehow knows but does not understand, is aware but has no control. The idea lives as a layering, a battle mask adhered, fused by fire to the face of life. It is a lie compelling to the patient's life in every way that matters, measured in the clinic as thoughts, mass, and actions. The doctor elicits and records anorexia's way of thinking, one of distorted self-image:

the patient states and believes one thing, while body mass index reports the opposite. The patient's actions too can be measured—reports on the restriction of food intake, which we can track as the patient does, rigorously counting all the tiny caloric ticktocks.

Immersive cognitive and behavioral therapies can help in anorexia nervosa—especially if prolonged, for months at a stretch—using words, and building insights, to slowly shift the distortions within the patient. The goal is to identify, and address, intertwined behavioral and cognitive and social factors, and to monitor nutrition with a touch of coercion. Medications are used not as cures, not to strike at the heart of the disease but to blunt symptoms; for example, serotonin-modulating drugs are typically brought in to treat depression that is often present. In some cases antipsychotic medications are provided that additionally target dopamine signals, and can favor reorganization of thinking, to help break the rigid loops and chains of the distortion; these agents can also cause weight gain, and so an otherwise-harmful side effect becomes a side benefit, to some extent.

There is much at risk. If including mortality from medical complications—the starvation-related organ failures—alongside the suicides, then eating disorders together show the highest death rates of any psychiatric disease. Decline and death come by failure of starving cells, all across the afflicted human body. Depression and suicide, if the first to fail is the brain. Infection, if the immune system falters. Cardiac arrest, if the electrical cells of the heart, already weakened from malnourishment, can no longer cope with the distorted salts of the blood—imbalances in levels of ions that had been set billions of years ago by the rocks dissolved in the ocean of our evolution, and now fishtailing free, diluted and fluctuating in the daily vagaries of starvation.

But for the survivors, the grip of the inner tyrant fades over time. The patient can writhe free, and impose by force new thinking and new action patterns—another layer of masking, perhaps, but still reaching at last a point, over years, when the story can be told like a nightmare.

·

Medications are just as off target for bulimia nervosa—which I suspected was Emily's secret—as for anorexia: able to blunt some comorbid symptoms, but still missing the heart of it all. Bulimia is also a killer with ion imbalance—wild swings in potassium and heart rhythm that come with the purge. Bulimia sometimes becomes mixed up with anorexia, as in Micah, together creating even more extreme shifts in fluids and charged particles—calcium and magnesium derangements too, in traces of rocks and metals needed to keep excitable tissues like heart and brain and muscle stable. These cells that twitch and spike need calcium and magnesium to function properly; otherwise spasms of spontaneous activity result: fibrillations in muscle, arrhythmias in the heart, and seizures in the brain—some ending in death.

The purge can come in many forms: self-induced vomiting, or laxatives, or even excessive exercise—anything that drives down mass balance. The mass balance credit is then used for intake—often with binges of food, piling the plate again and again, caloric reward multiplied by the illicit thrill of loopholing, from knowing the purge is coming, that nothing can stop its rush.

I knew that bulimia rush, that excited torture, from my time working with pediatric inpatients, and seeing it here in Emily I wanted to let her know I knew. If I were right, and if we could get it out into the open, together we could form a kind of partnership—a therapeutic alliance. From there, it would be a matter of logistics: starting some foundational therapy, building some insight, and discharging when ready to the right outpatient or residential program for her.

"Will you be able to tell us about it?" I asked, finally pressing. "I can tell you need to."

She was fully avoiding my gaze now, back to the bedspread. "I can't, really."

"Is it somehow related to why you can't stay in class?" I looked briefly over at Sonia the strong. She seemed enthralled.

"Yes, it is kind of the same."

Time to push a little harder; on the inpatient unit, we did not have the weeks or months of time that outpatient therapy would allow, and there were other patients too. "Emily, you mentioned earlier that a long time ago you sometimes would throw up after big meals." She had described this as remote, and minor, and not connected to her current symptoms; but now it made sense as a reason to leave class. "Is it possible that's happening again?" Her finger, which had been tracing infinities and parabolas on the bedspread, paused; her eyes remained on the bed, now fixed at a point, frozen.

"What would happen if you were alone, Emily?" I asked. She looked up at Sonia.

"I don't know," Emily said, to Sonia. "Maybe it would be okay. But probably not."

I let a few more beats go by, and shifted in my seat. Sonia picked up this call and responded. "Emily," she said, "do you want me to sit with you and talk? I think the doctor has to go see some other patients pretty soon."

"Sure, that sounds okay," she said. "It's no big deal." She sounded a bit diffident, but it was the biggest of deals; it seemed that Emily probably wanted to get better. Another page had come from ortho, I really did have to go, but I could leave Sonia behind to find out more, to work her new craft, its course now well defined. I buttoned up, fared them well, and shuffled out of the room. There was no rush now; time and space were needed for alliances to grow.

As I made my way to the orthopedic surgery unit, I pondered the contrasting appearances of Micah and Emily. Micah suffered both anorexia and bulimia, but his bulimia purge strategy involved not regurgitation but walking whenever he could: pacing, circling, and even surreptitiously clenching his leg muscles while seated—all forms of burning calories. A cryptic purge, subtle, not classic bulimia—and overall he seemed mostly dominated by anorexia. He was inward-directed, a tight little bundle of sticks.

Emily could hardly have been more different. She was strong, extroverted, energetic, at a perfectly healthy weight—though who knew, perhaps she swung from one disease to the other too. During our interview, she had mentioned some caloric restriction patterns from years earlier.

Was there shared biology, despite how different these two diseases, these two patients, seemed? Anorexia was a rigid accountant, tracking every calorie and every gram, suppressing the reward of food; bulimia was natural reward embraced, amplified, repeated furiously through a cloud of calories. Yet there was a paradoxical commonality—these two could still coexist, and even work together. Both were content to kill, but the compatibility seemed to me to be deeper yet; both achieved a toxic liberation, an expression of self as dominant over the self's needs.

What brain but of a human being could make such a thing happen? At what moment in evolution did the balance of power finally tip toward cognition becoming stronger than hunger? There was no way of knowing, but I guessed it could not have been long before we emerged, not long before we became modern humans. Wanting such a thing is not enough. Wanting to live beyond want—that is unremarkable, and universal. The hard part is making it happen, for anything as fundamental as feeding. But the modern human mind has vast and versatile reserves standing by, waiting to engage, to solve anything—calculus, poetry, space travel.

Motive force might be drawn from many different regions across the rich landscape of the human brain. Defiance of hunger is no small task, but for a nation of ninety billion cells it is perhaps not too hard to arouse powerful million-strong ensembles. Many different brain circuits could even individually suffice for the uprising, each in its own right a massive and well-connected neural structure, each adapting its own mechanisms, its own culture, its own strengths.

And so diverse paths might be taken to anorexia nervosa in different patients, depending on each individual's unique genetic and social environment—a complexity already hinted at by the diversity of genes that can be involved, as with many psychiatric disorders. One patient might raise an army against hunger by drafting circuits in the frontal cortex devoted to self-restraint; another might instead work through a self-taught cross-linking of deep pleasure circuits with survival-need circuits, learning to affix the attribute of pleasure onto hunger itself; still others, like Micah, with both bulimia and anorexia, working with both motion and thought, might find their way by recruiting rhythm-generating circuits, ancient oscillators in the striatum and midbrain built for repetitive behavioral cycles. Controlling the walking rhythms of the brainstem and spinal cord, via compulsive exercise, could suborn the pleasing rhythms of counting—for both steps and calories. With bulimia and anorexia, Micah would be counting both—the calories coming in and the steps going out, the tick and the tock. Micah had woven a soft repeating rhythm of the two, their rough-knit interlocked texture absorbing all his blood and salt.

Repetition is immensely compelling. Circuitry for repetitive grooming in birds—maintaining feathers in form for flight—does not need to provide awareness of any underlying rationale. Evolution just confers motivation, to loop the action without logic or understanding, front to back, again and again and again, pleasing and inexplicable. Or the digging behaviors of the ground squirrel, badger, and burrowing spider—each of these species locks the rhythm of the dig to its own specialized frequency, its tuned neural cycle from central pattern generators. Or scratching in mammals like us—every animal has a different dig—getting to the parasite and rooting it out, driven by the flush of reward that comes with the scratch as the itch is hit, once started barely stoppable, the rhythm only heightened in intensity by the necessary damage done to the skin. A full valence flip—raw pain now raw reward.

Our brains play out more complex rhythms too, spanning time and space, using the metaphor of these basic motor actions. The same frontal cortex that plans and guides our scratch with a hand, in lockstep with its deeper partner the striatum, is also an executive for planning the daily routines, the seasonal rituals, the yearly cycles. The reward of rhythm shows up across every timescale, and in nearly every human endeavor: in knitting and suturing, in music and math, in the conceptual rituals of planning and organization. Not only actions but repetitive thoughts too can become as compelling as any tic; the extension of ancient rhythms to new kinds of conceptual digging may help us build civilizations—but when the rhythms grow too strong, some of us become collateral damage: the obsessive-compulsive cleaners, the hyperaware counters, the groomers, the scrutineers, all the suffering relentless.

My pager went off again as I entered the ortho unit—it was the psychiatry housestaff office. I picked up a phone at the nearest nurses' station and called back. It was Sonia. "She's gone."

"Uh . . . what? Gone?"

"As soon as you left, she said she wanted to sketch her problem for me." Sonia's voice was tremulous, fear gasping out through every inter-syllable. "She asked me to get some markers, so I ran to the housestaff office, and came right back." She had imagined the thrill of diagnosis, maybe a publishable case report, an epic win for her residency inter-views. "I was only away thirty seconds, and when I came back she was just gone. She wasn't on a hold, so nobody was watching, and none of the nurses saw her leave."

"I'm coming back now," I said. "Sit tight, it's okay." But it was not okay. I had read her all wrong. Emily had been the cagiest of the psy-chotic depressed, suicidal but just wary enough to snooker me. In her own hidden craft, she had set out alone. The excitement of her final liberation was what I had been picking up, not knowing, misdiagnos-ing. My house of cards had come down, and I was responsible. I double-timed it back to the unit, just short of a run. Weak.

•

It was a complicated situation, but Sonia was right that we had no con-trol. Emily was eighteen and not on a legal hold. She had never ex-pressed suicidality, and had the freedom to come and go. There was no recourse.

We buzzed around the unit, looking for clues. She had taken noth-ing, and she had even left her laptop and phone exactly where I had seen them by the bed a few minutes earlier. Not what someone leaving against medical advice would typically do, if she were just headed back home or to a friend's house. There was no time, and no need, to speak our deepest fear.

We paged the attending to update him, though there would be noth-ing he could do either. It was on us, on me.

Only ten minutes had passed. The hospital was buttoned-down pretty tightly; even where there was no locked unit, the windows were generally sealed. If suicide were the goal, it wasn't clear what route she could take. We were on the second-floor open unit—I knew how to get to the roof of the fifth floor, through the hidden rathole of our resident gym, but there was no way she would find her way there.

Sharp edges . . . the hospital cafeteria, first floor, almost exactly under our feet? Or worse, just past the cafeteria, there was a balcony, an overlook into a vast atrium—it was a long way down from that balcony to the basement floor below. She could have gotten there in thirty seconds, and anything—everything—could have happened since then.

Sonia knew the stakes and was feeling it; her face was hard and I could see, just under the surface, cracking fault lines of failure and self-doubt. "Okay," I said, as reassuringly as I could. "She is probably just going to get a cigarette. In fact, that's probably what this whole thing is—the school stuff too." It was almost plausible—a flash recall brought me back to second-year residency, when I had been called in a panic by the labor and delivery unit; a new mother was demanding to leave right after her C-section and the whole floor was in a tumult. I had been called as the consult-liaison psychiatrist to—as the obstetrics resident put it—"*I don't know, place her on a hold or something.*" After speaking with the patient in her mother tongue for just ten minutes I got to the real reason—she only needed to go have a cigarette, and had been too embarrassed to ask. I savored that small victory for years, in part as crystallization of a curious and recurring lifelong theme—I had noticed that truth can always be discovered just by letting people speak.

But not this time, and that wasn't Emily. When you're just desperate to sneak off and smoke, you don't tell authority figures to sit with you. I kept that thought in, for now. "Hang on," I said to Sonia. "Let's split up. You go check the ER and the parking lot. I'll head toward the other side of the ground floor. Don't run." On task, Sonia, high ponytail describing frenetic horizontal figure eights in the air, was gone.

When she turned the corner, I speedwalked to the escalator, trying to project professional calm as I headed down to the ground floor. Ten seconds to the cafeteria, twenty to the atrium. I turned right, one more hallway to go. Counting steps. Listening for screams. The only sound a ticktock, each step a small victory, each step burns calories. Each step is a win. Nobody can stop you from taking more steps—and every step is closer to death.

I had been so close, but I had betrayed my undeserved gift, the inescapable theme of my life, that people seem to unburden themselves with me—and this time, someone needing help had started to connect, and I had walked away. Why? Only because orthopedic surgery had paged me once too often about a transfer that could wait.

Here. Sharp edges around this corner, at the sunlit cafeteria entrance. I allowed myself to think: it was a beautiful day, as was every day here. Sunlight was coming, but I was ready for that darkness, for that crow shadowbird.

The sun streamed in from the cafeteria patio as I turned right again, and there she was, just an arm's length in front of me: Emily. We nearly collided.

She had been intercepted, hustling out of the cafeteria entrance. Standing there, we locked eyes and then both looked down. She let herself giggle in relief. In her hand was a plate of food, piled high, nearly architecturally impossible. Fried chicken drumsticks, cake, pizza—an edifice of pure caloric reward.

She told me later it had been her third round-trip in ten minutes. Duck into the cafeteria, stack food, come back out through the entrance without paying—then to the patio to gorge, purge, and return. A cycle of reward and release, without consequence—yet hoping, needing, to be caught. Loopholing: victory over the body and the equations of mass balance. That was everything, and there was no stopping. But she felt it to be crazy, knew it to be dangerous, and did not want to be alone.

•

I was on call that night, and at the first quiet moment I went out alone onto the roof, through the door near the call rooms where residents could get a few minutes of sleep between admissions and consults, and onto the moonlit expanse of concrete and railings and vents. On rare quiet nights we would go out there together sometimes, two or three of us, residents or interns or students, and sit under the stars, leaning back against hard metal scaffolds in our thin scrubs.

The roof was uncomfortable but had the feeling of a sanctuary—a space apart before the next burst of calls and pages. That night it felt important to be still and alone, to consider what had happened with Emily. Something about the biology of this disordered eating felt hard and unallowable—and when that feeling comes, I have found, it is best to seek a moment to sit with the mystery.

This disorder seemed to me unique, and important, and a clue to something scientifically deep, but first I had to ask myself: how much of this strong reaction I was feeling—that neuroscience needed to learn much from this disease—was driven by my own parental sympathies, a drive to care for Emily, displaced? I relived another scene: of a father at his fourteen-year-old's bedside on the pediatric anorexia unit, in his oil-change-shop shirt—Nick, it says above the left pocket—after her heart attack and pneumothorax. The possibility of death has been spoken, and is known to him. He's no longer able to look at her; he is only just holding her, touch his only sense, seeing nothing, focusing only on the frail sparrowlike form of her scapula, her intermittent heartbeat felt faintly through to his chest, every two seconds, and her weak cool breath on his shoulder. No—he is remembering the sound of before she was born, the thumping whoosh-whoosh of her heart coming from the ultrasound like a war drum, filling the room, fierce and strong and fast, she couldn't be held back, she was his and coming, and the tears crashed out through his eyes, then and now. She was, she had to be always, unstoppable.

With the heels of my hands I squeezed my eyes and blinked at the moon. Here was the essential conflict I saw: the self was at war with its own needs.

To understand the biology of disordered eating, it seemed, we would have to understand something even more fundamental, and no more accessible—the biological basis of the self. And if the self could be separable from its needs, what was the self then? What lay within and without its boundaries? An ancient question, unsolved. We feel at home here—we are natives; we are the self, we think—and yet we cannot precisely draw our borders, nor name our capital. Not as human beings, not as neuroscientists, not even today.

Some bounds can be guessed. The self does not extend outside the skin, for example. But even that distinction is not as obvious as it seems. Parenting may seem to blur that line. Nor does the self fill the whole volume under the skin, nor even the whole brain. The self feels the body's needs—but these needs are broadcast by some agent that is other, yet still within the body. And pain or pleasure, doled out by some deep and dour neural banker—our suffering when drives are unfulfilled, our joy when drives are met—these seem only currencies that motivate the self to act, but are no more self than any monetary instrument: assets and debts, incentives.

Philosophy, psychiatry, psychology, law, religion: all have their own perspectives on the self. Just imagination, without exception, though each fantasy nevertheless describes truth of a sort. But neuroscience, with its power to know a new kind of truth, and to make that truth known, has not quite weighed in with an answer. Caution is needed—the right scientific words might not even exist yet. Perhaps there is no such thing as a self, after all.

We do feel at times an especially strong sense of self—for example, when we struggle with, and resist, and overcome a drive—but that sense of self could be illusory, and the victorious entity just a shifting alliance of other competing drives. Still, studying the process of resisting primary drives (with eating disorders as extreme examples) might be useful, since in late-stage anorexia, the entity that resists food is not obviously a rival drive. There seemed to me no clear natural process competing against the hunger—no reason to resist that the patients knew or understood or could express—and yet hunger could still be resisted. True, the resistance to food had started for a reason, as a primal drive—social pressure, leading to a weight-loss goal—but that was only the trigger, starting the conscription of cells and circuits into that vast new army, at the end with no more reason for its final ravaging of the body than the fact of its own existence. And yet in the magnitude of its blind destructive power perhaps deep biology was revealed—just as an earthquake exposes shattered strata that show how the earth was made, in the very act of breaking the earth itself.

Biologists speak of genetic mutations that are "gain of function" or "loss of function"—this means that a change has happened, a mutation, that turns the function of the gene up or down. These mutations help reveal what the gene is for. Knowing what happens with too much or too little of something reveals a great deal about that thing's role. For Micah's severe intake restriction, despite all he had lost, I could start with thinking of this behavior as a gain of function in the self, in whatever can resist the natural drive to eat when hungry or to drink when thirsty (of course with no implication that this self's distorted form is good for the human being, any more than a gain-of-function mutation of a gene is good, rather than destructive). But if one could eavesdrop on the activity of neurons throughout the brain, one could listen for, and localize, a circuit that stands out for resisting the impositions of the drive, at least under some conditions—and that could recruit allies, other circuits, to help suppress drive-satisfying actions.

Interesting enough for a starting point, I thought—and with a tractability lending itself well to exploration. But this starting point would be understood, from the beginning, to be only a simplification, since the self includes more abstract and complex representations of drive control than eating or drinking—extending to all principles and priorities, roles and values. And there was, I realized, a different dimension as well—also within the self, and helping to define the self, but wholly separate from priorities and primary drive adjudication. This separate dimension of the self was its memories.

Starting to feel the chill of night but unwilling to leave the moonlit roof just yet—the night having a perfection of its own, in a moment and a memory that might last—it seemed to me that memories of what we have felt, and have done, might be as important a part of the self, and just as fundamental, as priorities. If an external force were to change my memories, I might feel even more a loss of self than if my priorities were instead changed.

When answering which is the most important part of a self, it might matter who is asking.

On that rooftop, among the metal gantries and humming vents, when I thought about almost everyone else in their world—co-workers, leaders of society, strangers on the street—their priorities seemed a more important aspect of their selves than their memories. *More* important, actually, in that any change to those principles would matter more to me. The selves of others were in a different category, since for my own self, the opposite was true: memories mattered more than priorities. Loved ones were perhaps in between; my son's memories seemed as important as his priorities. A little blurring of self-boundaries, perhaps. Relationships extend the self into the world, through love.

Why do our memories, of personal past experiences, matter so much to our sense of self, with significance that is at least comparable to our principles? Since we don't control our memories, it is odd that we see them as essential to our selves—even the experiences that are clearly external and brought upon us, like a surprise kiss or a rogue wave.

In considering this puzzle, out alone under the just perceptible stars, a unifying answer began to emerge: perhaps our sense of self comes not from priorities alone, nor from memories alone, but from the two together defining our path through the world. The self could even be seen as identical to this path—not a path just through space, but through some higher-dimensional realm, through three dimensions of space, one of time, and perhaps a final dimension of value— that of worth or cost in the world, with valleys of reward and ridges of pain.

We are not defined by obstacles and passageways that were laid down by others, by nature, and by the body's inner drives. These details are not us. Other people, and storms, and needs come and go, and as they do, they alter the hills and valleys of the landscape—but the self chooses the route taken. Priorities pick the path. Our selves are not the contour of that landscape available to us, in this complex topography we travel—rather, they are the chosen path. And memories serve to mark the path along the way, so we can find ourselves, embodied as where we have passed.

In this way I could see self as the fusion of memories and principles, collapsed into the unitary element of path.

It wasn't clear how to make progress right away with any of this, and I didn't get far that night before the pager, once again, summoned me to the hospital below. Though I asked myself these questions all the way through training, it took fifteen years from the day I met Emily for neuroscience to respond to me, to say anything in return at all. And when the words of science were spoken on this at last, they were consummatory, spoken in the tongue of intake: of food and water, of hunger and thirst.

·

Milton's fallen angel in *Paradise Lost* considered worldly losses as trivial compared with the stability and surety of self, of *a mind not to be changed by place or time*—even when freshly fallen into hell, a setting known to eating disorder patients and their families. Most of us are familiar with, and have used from time to time, this psychological defense. *Here at least we shall be free:* suffering is tolerable if it is the price of freedom.

This perspective helps define the self in a useful way, as that which will accept suffering rather than serve the tyranny of needs and comforts. The self makes, and is, its own place in space and time: defined not by need or circumstance, but by choosing a path that resists need. What then are the brain cells and regions that might have the capability, and agency, to pick such a path: to define a trajectory through the world that resists intense need (without just satisfying another drive)? Such circuitry would bring about a special kind of freedom, and in some patients enable a special kind of hell. Neuroscience has recently brought just a crack of light to this problem, illuminating the line between need and self, edging open this door to mystery.

Hunger and thirst, two of the most powerful drivers of animal action, begin in the brain as neural signals from small but potent populations of neurons deep in the brain, cells mixed together in a dense jumble with diverse roles that seem unrelated to one another, in and around a structure called the hypothalamus. The hypothalamus lies deep; the prefix *hypo* reflects its steadily progressing evolutionary burial *under* eons of neural sediments—under the larger thalamus, which itself is under the much larger striatum, which in turn lies under the most recently laid-down cortex forming the densely woven neural fabric on the surface of our brains.

Some of the first optogenetic experiments were carried out in these depths—in fact, the first optogenetic control of free mammalian behavior was in the hypothalamus. In 2007 only one kind of neuron here—the hypocretin cell population—was made responsive to light delivered through a fiberoptic. Control of waking and sleeping, and the REM of dreams, resulted; providing millisecond-scale pulses of blue light, twenty times per second, to these specific cells in this part of the hypothalamus caused sleeping mice, even in REM sleep, to awaken earlier than they would have otherwise.

This new precision had been needed, here as much as anywhere in the brain—since the hypothalamus holds, within its seemingly chaotic mix, not just neurons involved in sleep but also cells for sex, aggression, and body temperature, as well as hunger and thirst, and virtually every primal survival drive. All of these cells serve as broadcasters of individual needs—imposing (or trying to impose) their message on the broader brain, on the self wherever it may lie, to drive action addressing that need, working the levers of suffering and joy as needed to reinforce that action. But all of these cells are intertwined with one another in the hypothalamus, not separately accessible in real time by scientists seeking to test for roles in behavior.

Yet with optogenetics, gain- or loss-of-function experiments were enabled, to reveal how primal survival drives arise from specific activity patterns in single kinds of cells—or even single cells. Neuroscientists could control—provide or take away—the electrical activity of any one of these diverse types of intermingled cells selectively, using the same optogenetic principle that had illuminated anxiety, and motivation, and social behavior, and sleep: in which genes from microorganisms cause production of light-activated electrical currents only in the cells of interest.

Optogenetics allowed testing which of these deeply buried hypothalamic cells—known to be naturally active during the state of need—in fact cause the hunger or thirst behaviors, actually driving the consumption of food or water. Spots of laser light were delivered into the brain by fiberoptics to turn on or off targeted cell types in the hypothalamic region as animal actions were chosen in the world. With the flick of a switch to drive optogenetic excitation, a food-sated mouse immediately began to eat voraciously, and the opposite experiment—an inhibitory optogenetic intervention—suppressed food consumption even in a hungry mouse, underscoring the natural importance of these cells.

Similar experiments were carried out on different hypothalamic cells: those for thirst. These experiments showed, in the starkest of ways, how the choices of action made by an animal can be determined by electrical activity in certain very specific, and very few, neurons deep in the middle of the brain. The conundrum of agency (does free will meaningfully exist or not?), while not answered, is particularly well framed here. That a few spikes of electrical activity in a few cells control choices and actions of the individual—this now cannot be denied.

Watching these effects in real time in mice, a psychiatrist can become awash in personal memories—heartrending clinical images of bulimia and anorexia, seeing an individual gorging on food not needed, or suppressing food intake desperately needed. The hunger and thirst optogenetics experiments provided proof of principle that a local cluster of cells deep in the brain could cause and suppress such symptoms—and so, perhaps, that we might be able to design medicines or other treatments to target these cells.

But there was a key difference between experiments with optogenetics and realities of the disease—a distinction important for treatment, and for understanding the basic science of the self. In the optogenetics experiments, we directly accessed—turning up or down—the deep need cells that broadcast the drives of thirst or hunger. But bulimia and anorexia patients, despite their extreme thoughts and actions, still know the hunger—or at least the emptiness—is there. What the patients may do is counteract the effects of that feeling—associating positivity with the emptiness. If patients can't get to the need cells in the hypothalamus directly—outside the self's conscious control—this is how it must be done. Bring opposing resources to bear, fighting the effects of those need cells, forming a large enough and strong enough crowd in the town hall to win, to outshout hunger.

Is this how anorexia and bulimia become endowed with personhood? Riding atop the self circuitry, while clearly separate—a parasite, a virus recruiting the machinery of the host cell, a shell running atop an operating system, an emulation of a self. Only in this way can the disease access the problem-solving capability of the human mind. The disease recruits all the brain that the self normally can access, and must access—by turning hunger into a problem to be solved.

This simple subversion, initially endorsed by the patient—turning hunger into a challenge—allows recruitment of what our brains seem so well evolved for: solving problems, in a general and abstract way, to address needs that could never have been anticipated by evolution. And perhaps if we were not such versatile problem solvers, we would never have developed the ability to suffer this class of disease. As I had considered the day we lost—and found—Emily, different patients may solve the problem with different tricks—some by using circuits that are expert in discrete repeated actions like the striatum (to bring in the OCD-like pleasure of the rhythms of the count, and strike, and dig, and scratch, and weave), perhaps others by using restraint drives located in the frontal cortex (to bring in powerful executive-function circuits that suppress feeding in the context of social cues).

These are intriguing possibilities, but not far-fetched; in 2019, optogenetics experiments directly revealed groups of individual cells in the frontal cortex that were naturally active during social interaction but not feeding—and when directly activated by optogenetics, these specific social cells could suppress feeding, driving a resistance, even in naturally hungry mice. But regardless of specific provenance for one patient or another, the militia called up are strong circuits, and expansive, even though some have only emerged recently in evolutionary time, like those of the neocortex—that thin and vast sheet of cells including frontal cortex, a problem solver partnering with the deeper and older striatum, its enforcer and link to action.

Rodents have much smaller brains than we do, and relatively less neocortex, so mice may be less suited to resist drives. But neocortex they do have, and in a separate stream of optogenetics experiments from 2019, it was discovered that certain parts of the neocortex can stand apart from even strongly driven primal drives. When a mouse is fully water-sated but the deep thirst neurons are optogenetically driven, intense water-seeking behavior results—and yet a few parts of the brain are not fooled, and seem to know the animal is not truly thirsty. These circuits listen to the drive, without buying in; their local neural activity patterns are only mildly affected. This result was one of several discoveries resulting from the kind of brainwide listening experiment I had hoped for years earlier: using long electrodes to listen in on tens of thousands of individual neurons across the brain, while optogenetically stimulating the deep thirst neurons.

The first important finding from this brainwide eavesdropping, and a big surprise, was that most of the brain—including those sectors thought to be primarily sensory, or just movement related, or neither—was actively engaged in the simple state of seeking water when thirsty. Perhaps this finding revealed an important natural process by which the brain keeps all parts of itself informed on all planned movements and goals, so that even simple actions would be experienced by every part of the brain as generated from the self, and there would be no confusion as to the source of the drive for action. This unitary quality can go awry in disorders like schizophrenia, where simple actions can feel alien—as if generated from outside the self.

More than half of all the neurons recorded across the brain showed engagement with the task of acquiring water—both when the animal was truly needing water and also when, with optogenetics, we created a thirst-like state. So now not only are those old tales (usually thought to be false) claiming that only half, or even 10 percent, of our brains are used for this or that, demonstrably wrong—but it also seems likely that almost the entire brain is activated in specific patterns during every specific experience or action (since we now know a task as simple as drinking water when thirsty involves most of the neurons throughout much of the brain).

The second key finding was localization of the resistance: that identification of brain regions refusing to be intimidated by the imposed deep drive. Although clearly affected and so undeniably hearing the thirst signal sent from below, those few cortical structures, recently evolved and at the surface of the brain, stood out. These were not fully responding, not matching the state they would have entered into for an animal naturally seeking water when thirsty. The resistance was revealed like a shadow thrown, across both the *prefrontal cortex* (a region already known to be responsible for the generation of plans or paths through the world, and locating oneself on those paths) and the *retrosplenial cortex* (a region already known to be tightly linked to the entorhinal cortex and hippocampus, two structures involved in navigation and memory of paths in space and time). Both prefrontal and retrosplenial cortex thus fit the idea of self as path, and were already well known to be active during stimulus-independent thought—when a human subject is asked to sit quietly and think of nothing in particular, to simply be with one's self. This pattern contrasted with that in neighboring cortical areas (insular cortex, anterior cingulate cortex, and others), which showed neural activity patterns nearly indistinguishable from when the mouse truly needed water, when thirst was real.

So it seems many brain areas are able to feel and encode the state of natural thirst, as they should in order to help guide appropriate action to keep the animal alive. But at least two—prefrontal and retrosplenial cortex, perhaps in their role of self- (or path-) creation and navigation—in some sense know more about what the priorities of the animal should be, in terms of where it has been and is going, separate from the deep thirst drive. These two regions reside in recently evolved brain regions—quintessentially mammalian, and massively expanded in our lineage.

It is on the backs of such resistance that the eating disorders may find their strength—a standing army, quartered in neural barracks but always restless and poised to be called up by disease. Like the self-circuitry I imagined years earlier on the frigid metal gantries of the moonlit rooftop—thinking of, and recovering from, Emily—these parts can come to war with the whole, and win.

•

I walked Emily from the cafeteria to her room—she was relieved to come back. We lined up staff members to sit with her, which required some negotiation; there was no compelling legal authority against bingeing and purging, though we had some leverage since she had been stealing food. Sonia stayed with her first, Sonia who had been transformed back to her old self, with all her strength and even her serenity restored. And Emily at last could rest, cut off from access to the actions of bulimia for the time being; she could begin to recover and participate in developing a long-term plan for full healing. Even as we worked to make sure Emily was not left alone, our social worker began charting the path to an outpatient program. Bulimia had not dwelt long in Emily, and we were hopeful when she was discharged two days later.

For Micah, who was in his forties with behavior that seemed so entrenched, I was much less sanguine. We had already tried everything at our disposal. We could continue to occasionally place the nasogastric tube for feeding, when his blood pressure and heart rate skewed dangerously low—but the legal basis for doing this was always shaky and depended upon his inconsistent consent. He was not suicidal or homicidal, which is all the law addresses in allowing compulsion of psychiatry-based treatment—either these or grave disability, the inability to provide for his own basic needs. But Micah could provide perfectly well for his own needs; he only chose not to. Doctors can also compel emergency care if the patient is unable to understand the nature and consequences of treatment and cannot make an informed decision—but here again, Micah understood perfectly all the choices and consequences. He was not in delirium or psychotic. He simply wished his body to take a certain unusual form—with all the attendant risks. Here at least, he could be free.

As Micah continued to occasionally accept the NG tube— apparently only to toy with me, removing it himself later at night— I wondered how I seemed to him during all this. Hapless and childlike, arrogant and threatening—or more likely, I was not even worth that much thought. Micah's double disease set a course for him that was so strongly overdetermined he could chart his own path up the steepest hills of pain in that realm of space and time and value, and anything I said or did was beneath notice, just a slight shift of gravel under his feet. He refused a last-ditch medication that we hoped could help organize his thoughts, a low dose of olanzapine—which we also thought would put some weight on him as a side effect. I rotated off the service a week later, leaving Sonia to work with Micah. He was discharged to an outpatient facility a few days later, no better for all our ministrations.

•

Sonia collapsed during a psychiatry team dinner at another resident's apartment later that month. I hadn't seen her for three weeks. David, a neurosurgery resident, partner to another psychiatrist there, was standing next to her and leapt into action. Sonia had not quite lost consciousness, but David did a quick check there on the carpet, and then a fuller exam after we brought her to the couch and gave her orange juice. We hung back and let him work her up, until he finally stepped back, satisfied she had only fainted and was stable—and then, in a surreal moment, apparently because I knew Sonia best, David came over to present the case to me as if I were the attending rather than just another resident like him.

As worried as I was, and wanting to talk quietly with her myself, I remember thinking in that dimly lit room what an elegant thing his presentation was. David ticked through the history he had acquired, summarizing the medical and neurological exam he had tapped out in the miraculously intimate sonar of the physician without instruments, the pianist-like rhythmic fingertip percussing of internal air and water and organs, of reflexes, of heart rate and blood pressure—and concluded Sonia was severely dehydrated. She had been working out hard, admitting to eight- or nine-mile runs every morning and eating little—just not enough time, she had said. That day, there had been just carrots and some coffee.

Trying to peer around David, I did my best to see Sonia through the gloom as she lay on the couch across the room. She looked the same as when we had been on the team together, not thin or weak. What had I missed, then, about Sonia the strong? Or perhaps instead, she had only recently come to this way of being, in the past few weeks having been joined by another, now sharing her journey.

If anybody could solve the equations of mass balance and create a path, a state in defiance of a primary drive, it would be Sonia. She was her movement, she was her path, and there can be no self without movement along the path. Resist? One may as well. She had that part that moves, and fights back, and for it goes to hell.

CHAPTER 7

MORO

The broken dike, the levee washed away,
The good fields flooded and the cattle drowned,
Estranged and treacherous all the faithful ground,
And nothing left but floating disarray
Of tree and home uprooted,—was this the day
Man dropped upon his shadow without a sound
And died, having labored well and having found
His burden heavier than a quilt of clay?
No, no. I saw him when the sun had set
In water, leaning on his single oar
Above his garden faintly glimmering yet . . .
There bulked the plough, here washed the updrifted weeds . . .
And scull across his roof and make for shore,
With twisted face and pocket full of seeds.

—Edna St. Vincent Millay, "Epitaph for the Race of Man"

"Mr. Norman, he's on 4A. Eighty-year-old veteran, long history of multi-infarct dementia. Brought to emergency yesterday, by family." The medicine resident's voice over the phone was pressured—getting his work done, trying to check off this consult request as quickly as possible. "They report that he slowly stopped talking, progressing to total silence over the course of a couple months. That's the only new symptom."

In my mind, that history was already concerning for neurological disease, raising the specter of a new stroke—especially with the apparent history of brain infarcts in the past—but it would be odd for a stroke-related process to evolve over months this way. I noticed in myself a mildly rewarding feeling of intrigue—a sensation I recalled from chess, when encountering an unconventional opening move. It was such a pleasant sensation, I even felt a bit guilty for feeling it. I leaned back in my chair and looked up at the grimy, flaking ceiling of the hospital sandwich shop. "Interesting," I began to reply, only to be cut off as the resident brusquely continued.

"The patient just moved here from Seattle after his wife died," he said. "Been living with his son's family in Modesto for a few months now. The family was worried about another stroke but we saw nothing new on the scan last night—just old white-matter lesions. He did have a UTI, so we're treating it, and we admitted him last night for that and to work up whatever happened with his talking. And now, guess what?"

A pause for effect—despite the pace of his pressured speech, the resident could not hide that he had found this one interesting too. Moments of intellectual reward can be frustratingly short-lived on inpatient call shifts, with little time to satisfy human curiosity; here, for what it was worth, one such moment seemed to have come.

"I got him to talk," the resident continued. "Turns out he can when he wants to. Just a real unpleasant character—he doesn't care about anyone, didn't care that his family was worried. Extremely cold, actually. I think antisocial personality. Guess even you guys can't unfuck that." Riffling sounds. "Still in the process of getting his records from Seattle, but their little clinic is closed till Monday. His son's here but doesn't know much of the medical history; they weren't a close family. Not a big surprise. My attending wanted me to call you, to see if you could evaluate for psychiatric explanations, since we can't find anything else. I don't really think it's delirium, since he seems oriented, though you could still suggest a trial of haloperidol—his QTc is 520, so let's be careful. Anyway, I think he just doesn't like people. This should be quick."

The resident had thought about side effects on heart rhythms, and rightly so—if the interval between two peaks on the electrocardiogram was already as long as 520 milliseconds, the treatment team risked causing a serious heart arrythmia with certain medications such as haloperidol—but his antisocial personality disorder idea didn't sound right to me, and diagnoses I favored more began volunteering themselves, populating my mental workspace. It was more likely, I thought, to be a form of delirium that did not fit the resident's expectations—

a quiet subtype of the waxing-and-waning disorientation often seen in the elderly, sometimes arising as a medication side effect, or caused by a moderate illness like his urinary tract infection. The medicine team might have assessed him by chance during a lucid phase of the delirium cycle, and so found him to be oriented.

The quiet ones are often missed; many doctors expect a highly active, vocal, demonstrative state of delirium, but the presentation we call hypoactive delirium is one of withdrawal, silence, and stillness on the outside—while deep inside, a storm of confusion churns.

On the other hand, if the resident had been partly right—in the sense that there was not delirium but rather a personality issue—then the personality change that comes with dementia was more likely to be relevant here than antisocial personality disorder. That empathy-deficient trait of antisocial personality would have been part of a life-long pattern—and though unpleasant, would not have struck the family as unusual now. Also favoring the dementia explanation, our brain imaging had apparently confirmed the underlying process: blood flow blockages (lasting long enough to kill cells) in vessels that supply the depths of the brain with sugar and oxygen.

These infarcts, spots of dead tissue that are the outcome of strokes, can be detected by computed tomography even years after the blockages, as scattered holes in the dense silkwork of interconnecting fibers that link brain cells across long distances—showing up on CT scans as lakelike black gaps called lacunae. Even in patients without a known stroke, more sensitive technologies like magnetic resonance imaging (MRI) can show the small vessel blockade of vascular dementia in a different way, as a profusion of intense white spots—scattered across the brain, marking the end of day with light, like stars in the early evening.

Personality change in dementia—well, common things are common. These changes appear in all the dementia syndromes, along the way and especially toward the end, when individual parts of the brain that manage predilections and values begin to break down. I had seen Alzheimer's disease patients with newly aggressive—even explosive—anger syndromes; Parkinson's disease patients with sudden risk-seeking tendencies; and fronto-temporal dementia patients with almost-infantile self-centeredness, verging on antisocial behavior, that the resident might have sensed.

In dementia, memory loss is the most widely recognized symptom, but dementia does not mean just amnesia. Rather more fundamentally, the word means loss of the mind itself. Memories—the stored senses and feelings and knowledge from along the journey of life that infuse the path with color and meaning—are effaced along with the values that set the path's boundaries and direction. And the latter—changing personalities and upended value systems—can be as shocking as the memory loss: a fundamental transformation in the identity, the essence of the self, of the person known and depended upon for so long.

This was, I thought, the more plausible syndrome. But without seeing the patient I could not be sure; there was also a chance the resident had actually nailed the full diagnosis—perhaps a well-disguised antisocial personality disorder had been unmasked by another process, like his urinary tract infection. I began imagining the distinct chill of the antisocial, and reflexively steeled myself to prepare for that slick indifference, that simulation of social grace, that viperish gaze unwittingly revealing to me how little I mattered, showing that one cannot conceal what one does not understand.

It was a quiet late-spring Saturday afternoon, the regular weekday psychiatry consult team was off, and I was the on-call resident for all things psychiatry. It was on me, so I stood up from my small table in the cramped hospital café, donned my armor—starched white coat, stethoscope, reflex hammer, pen—cleared away my coffee cup, and headed to the fourth-floor medicine inpatient unit.

•

Each major medical specialty in the hospital provides an on-call consult service to help fellow physicians with complex cases. In psychiatry this service is called the consult-liaison team, and training in psychiatry involves a heavy dose of C-L, fielding calls across the hospital—from intensive care and medicine units for managing delirium, from the obstetrics floor for evaluating postpartum psychosis, from surgery for sorting out competency and consent issues, and sometimes just for patient transfer when a unit with a truly closed and lockable door is needed.

Highly interdisciplinary or mysterious cases, pan-consulted, can bring the hospital together—in a kind of block party of clinical care, with many services buzzing around. This had not been obviously one of those cases—with its seeming simplicity—though when I pulled the chart from the rack at the nurses' station, I discovered that several consult teams had already been called before me—the neurology service most recently. I was the last resort for Mr. N. (as the notes referred to him, in the veterans hospital culture of anonymous respect).

Among the possibilities not mentioned by the resident, but discussed in chart notes left by the various teams, were forms of parkinsonism; the speech therapy team had correctly noted that Parkinson's disease can involve slow movements and reduced vocalization. The neurology consult team, ultimate arbiters of Parkinson's, had then come and gone, confirming poor short-term memory and multi-infarct dementia—but they found no signs of Parkinson's, noting as they signed off that although Mr. N. never smiled spontaneously, he could move his facial muscles when requested; this was not the frozen, masklike state of parkinsonism.

Neurology had commented also on the brain imaging confirmation of his multi-infarct dementia; recent and distant strokes look very different with these scans, and since no new stroke was apparent with CT, Mr. N.'s new reluctance to speak needed some other explanation. So psychiatry was called last—completing the usual progression through medical specialties, ending in the realm of the unknown.

I found Mr. N. in bed, looking straight ahead, and oddly still. His bristly bald head was propped up on three pillows, and his corrugated cheeks seemed to glisten slightly under the fluorescent lights. I also thought this was not Parkinson's after my own physical exam—there was no parkinsonian rigidity of the limbs and no tremor. Nor did I see signs of catatonia, a rare syndrome of immobility I needed to rule out, which can arise from psychosis or depression; he could move all his muscles readily when asked, nerve by nerve.

Delirium could also be mostly excluded—with the unlikely caveat that this might be, by chance, just another lucid moment. As the medicine resident had said, Mr. N. could speak, and he spoke a few words to me, choosing to answer only when asked repeatedly and the question was simply factual—but this was enough to establish that he was mostly oriented to time and place. Mr. N. knew he was in a hospital, he knew who the president was, he even knew the state we were in. He knew his son's name—Adam, from Modesto—the one who had taken him to the hospital this time, the one who had brought two grandchildren into Mr. N.'s life.

Though he refused to answer questions on his internal state—by remaining impassive, or with a brief head shake—one of his refusals came with a subtle feature that I could have easily missed had I not been watching carefully. As part of the full mental status examination in psychiatry, we probe participation in everyday interests and hobbies—asking if these are being pursued and enjoyed. The question sounds conversational, but reveals a great deal about motivation, and the ability to feel pleasure. His response to this, my query if he was enjoying his normal interests and activities in life, was nonverbal—just a downward twitch of one corner of his mouth in a hint of grimace—a half beat of self-disgust that seemed to me incompatible with delirium or antisocial personality.

I then had an urgent responsibility, one neither the medicine resident nor I had anticipated. Having caught a glimpse of his inner state, I now had to rule out depression, perhaps with an accompanying paranoia (this can be caused by severe depression and could explain his reticence to speak)—and I had somehow to address this life-threatening possibility in a mostly nonverbal patient, despite the fact that every diagnostic criterion in psychiatry is ultimately verbal in nature.

If Mr. N. were traveling deeper into a storm of psychotic depression, more and more stoic on the outside as he became increasingly paralyzed internally by hallucinations and paranoia, this syndrome would be a disaster to miss—especially since the condition would be elegantly treatable, with straightforward medication strategies. Alternatively, even if there were no psychosis, but only a severely depressed state suppressing effort allocation—making it too much of a motivational challenge to articulate words, to move lips and tongue and diaphragm enough to maintain a simple conversation—this state would have to be ruled out as well. Such a severe nonpsychotic depression could be fatal, but would also certainly be treatable.

I needed an approach that did not require the patient to form words. Seeing a framed photograph at his bedside—a Modesto High School basketball player, she seemed to be maybe fifteen—probably left by his son, I asked Mr. N. to show me a picture of his granddaughter. Displaying no grandfatherly excitement or pride, just shouldering the burden of my request, he complied—but he had no interest in looking himself. He just directed me with his eyes toward the evidence—and was done. No hint of the disorganization of psychosis.

I picked up the photo and showed it to him, pointing to her, asking for her name, watching closely. There was not a twitch of a smile, nor a softening of the eyes; but his gaze was not as dry as it had seemed. Up close I could track the almost imperceptible glistening of his cheeks; I had thought it the faintest sheen of perspiration, but the hospital room was chilly and now I could divine the source, track its scattered and discontinuous path up through crevasses and forks to headwaters in the corners of his eyes. He remained quiet, and could not say her name. Silence crashed around us—deafening, negative, noise.

•

In major depressive disorder, loss of pleasure is a classical symptom, and a classical-sounding name is given: anhedonia, the absence of beauty and joy from life. As cleanly and completely as senses of taste and smell are lost during the common cold, pleasure can be somehow detached from experience.

Though I had seen the anhedonia of depression—this inability to find reward or motivation in natural joys—many times before, it was unsettling every time. I could see how the resident had been guided down the wrong diagnostic path. Such a symptom could have appeared to manifest as a sort of inhumanity to physicians, friends and family— with a seemingly reptilian lack of warmth, even for his own grandchild.

How many millions of people with depression, over the course of human history, have had their isolation and suffering compounded this way, by helplessly eliciting anger and frustration in others—exacerbating all the other challenges and agonies of their disease? Even with this perspective, I still had to work on my own cognitions, to not react negatively to him as a person. Knowing is one thing, but understanding is another. I knew but still didn't understand, not deeply, neither as a human animal nor as a scientist.

To understand how pleasure can be detached from human experiences so universal and fundamental, we might begin by asking how value is linked to experience in the first place—where and why, in the human brain? And where and why, in the story of humanity? The answers, if we could find them, might explain the fragility of joy.

Sometimes the allocation of joy is automatic. We can feel powerful innate rewards, which serve as natural reinforcers of behavior important for survival and reproduction. One of these preset rewards may be pleasure at interacting with a grandchild—an experience that seems naturally positive in valence for us, though heightened further by experience. This response (not universal among mammals) might have acquired survival value in our lineage only as primates became more long-lived and social, for its utility in encouraging protection and education of the young. Those with increased ability to link reward circuitry to representations of the extended family could have benefited enormously from such an inborn wiring innovation. But all such connections, as physical structures, are vulnerable like any other part of the brain to a stroke—and depending on the exact location of the infarct, the effect could appear to be either specific to one kind of reward and motivation (causing an upheaval in priorities, and so an apparent personality change) or a more general and pervasive loss of pleasure in life (like the nonspecific anhedonia of depression).

Other innate pleasures seem to make little evolutionary sense— their existence only underscoring our ignorance. The reward of seeing the wild rough seashore—without the promise of food, water, or company—does not explain itself well. It is not the joy of returning home, not as we know it, not even in an evolutionary sense. Our fish-like forebears learned to breathe at the brink of land and water, but not at the pounding interface of cliff and wave. That part of our story is more in the shallow swamps of 350 million years ago, when the first air-breathing fishes emerged onto land.

Why, then, do nearly all of us see beauty in the seashore? Is there innate intrigue in the stark contrast of cliff and crashing wave, the power and peril of momentum versus bulwark? Or perhaps the waves in some way evoke the windsway of an arboreal canopy, or the reliable repetition of a lullaby, soothing with its rhythm and inevitability. Whatever its meaning, this joy is real. It is widely shared, and runs deep, and yet no logic seems fully explanatory. There are many such examples.

Natural selection provides one potential answer for the meaning of joy, which is that there is none. Meaning is an elusive, even absurd quantity in evolution; there was no meaning underlying the emergence of mammals to dominate the world after the dinosaurs—it was just chance, a giant meteor strike compounded by other natural disasters sixty-five million years ago that killed off most life as ejected dust blocked out the sun. It was meaningless, but consequential, how suddenly valuable it was to be small, fast-breeding, warm-blooded, fur-covered—and to have a strong innate drive to live in holes.

Some feelings, and the resulting behavioral drives, might arise from such chance associations, just vagaries of the environment. If a small group of human ancestors had a spontaneous affinity for—and crafted their lives around—the seashore, then the unrelated bottleneck contraction of human populations many tens of thousands of years ago could have created a founder effect: a small set of survivors exerting a large effect on the subsequent population. If most of those human beings who survived hung on via mussels and tide pool debris, scraping along like limpets on wet rocks as the rich plant life and large game animals on land died off, surviving humanity might carry within a joy and an affinity for the seashore, an intense appreciation for its imagined singular beauty—a joy not caused by the population crash but simply allowed to persist and propagate for the time being, due to humanity's close call with extinction. Not to say we know anything like that happened—though from paleogenetics we can see that there were indeed bottlenecks for us, including that global collapse in human populations bottoming out only fifty thousand years ago. Our most mysterious instinctual impressions of loveliness, then, may be just accidental fingerprints—left by artists of survival, on the cave wall of our genome.

When we all do feel joy or reward without learning, this is a trace of the past, cast forward over millennia of experience in our lineage; our forebears, at some point, most likely felt that joy, and those who were able to feel this way were able to create us. But learned rewards are another matter, arising within a lifetime, even within a minute. The brain seems designed to ingest new information, and to swiftly alter itself in response—this is how memories are made, and behaviors learned or changed in the life of an individual—and these fast physical changes can be studied in the laboratory, providing a short-timescale model for what evolution might work with on longer timescales. Learned behaviors can be rapidly tuned by modulating the strength of connections in the brain, and the groundwork for innate reward-seeking behaviors might be set over millennia in a similar way—as evolved and genetically prescribed connection strengths inside the brain. Whether learned or innate, feelings may be attached (or detached) from experience via the physical expedient of changing the strength of certain connections across the brain. And so two distinct concepts—feeling and memory—powerfully converge, both in health and in their disordered states: anhedonia and dementia.

•

We needed Mr. N.'s medical records to see if depression had been detected before, if any hints of psychosis or catatonia had been observed, and if any psychiatric treatments had been tried—with success or failure or side effects. These data points could be essential in finding a safe medication and avoiding harmful attempts at treatment (an especially important consideration in geriatric psychiatry).

The Seattle clinic was closed until Monday, the resident had said, and it was still only Saturday night. I needed that information before suggesting a medication. The next step was for me to connect with the primary treatment team and put together a plan—but it was getting late; time for Mr. N. to sleep. For the moment he was stable and safe, and so I took my leave, letting him know I'd come back to him tomorrow with a plan. He did not respond.

As I reached and opened the door, already looking out into the hallway, I heard a voice behind me:

"It's going to be a long night."

I froze at the door. Unprompted, a full sentence had come forth—from this patient who had not spoken on his own at all before, and with only one or two syllables at a time when pressed.

I turned and looked back across the room. He was now eerily upright and looking straight at me. The glisten on his cheeks was more intense, only on his upper cheeks, near the inner corners of his eyes. The room dropped away. I saw him fully—his veiny, bald head rocking gently with each breath, the symmetric sag of his eyes and mouth, his steady gaze on me. He did not speak again. He had said something important that he needed me to know.

After a long pause, I gave him my warmest smile and a reassuring nod. "Don't worry, Mr. Norman, we'll be with you all the way through."

It's going to be a long night. The last sentence he would ever speak.

•

The long course of dementia—whether spanning years or decades—is almost certainly a new phenomenon of life on earth, willed into existence by modern medicine and effective extended-family care. Via supportive social structures built with our brains, we have made the persistence of dementia possible, and have not yet found a solution. There are no cures, and the few medications available only slightly slow the steady progression of the disease.

In psychiatry, dementia is today (and this will change again) called major neurocognitive disorder, which for diagnosis requires the conjunction of both a loss of independent functioning and a loss of cognition (which can include nearly anything related to memory, language, social/perceptual/motor function, attention, planning, or decision-making). This long list, and the diversity of the symptoms that are permitted for making the diagnosis, in turn allow dementia—or major neurocognitive disorder as a medical construct—to encompass all the small and large disruptions in brain communication that can occur over a lifetime: by lacunae from strokes, by plaques and tangles in Alzheimer's disease, by spots of focal damage from accumulated injuries.

Disconnection, miscommunication, lost pathways. But what is actually missing?

Although brain cells certainly die in the dementias, it is not known if memory loss is always due to something like the loss of cells or synapses responsible for holding the memories—akin to wiping a computer drive. It is possible that instead, for at least some stages of white-matter damage like multi-infarct dementia, memories remain intact—but lie isolated from input or output projections, with only their connectivity lost.

With only interrupted input—access to the memory lost, just the pointer or lookup information—the memory would be present but not reactivatable. Or perhaps only interrupted output could occur: memories might be reactivated perfectly well, but find themselves unable to reach back into the conscious mind. Slumbering in the snow, or screaming into the void—either way, a memory could still live intact but in isolation, with connectivity lost due to the dark lakelets, the lacunae, the focal infarcts severing long-range fibers spanning the brain.

Clinically, a substantial fraction of multi-infarct dementia patients also exhibit anhedonia—a surprising correlation for two seemingly unrelated syndromes. Studies have found considerably increased anhedonia in senior populations with cognitive impairment, versus cognitively intact comparison groups—even manyfold increased in patients with frank dementia. This connection between feeling and memory runs yet deeper; in these populations, the greater the accumulated volume of those lacunae in white matter—showing greater loss of long-range connections, the carriers and controllers of information—the more anhedonia was seen. When memory fails, feeling can follow.

Optogenetics experiments have shown that value, or valence as we say, can be attached to brain states by long-range connections across the brain—for example, that valence of release from anxiety is set in part by projections from the BNST to reward circuitry deep in the midbrain. And these intriguing human epidemiological linkages—the association between anhedonia and dementia, and the association in dementia between lacunar volume and anhedonia—could be explained if the same process that causes the decline in memory (damage to long-range white-matter tracts, inputs and outputs) also causes the decline in feeling. Cells that could provide feelings may still be present but cut off, in the same ways that memories may be lost: by becoming voiceless.

Memory also needs feeling, in some sense. There may be little justification to store and recall a memory of an experience, unless the experience matters enough to elicit a feeling. Information storage takes space, uses energy, and creates curational challenges; no such cost can be borne for long on evolutionary timescales without manifesting some benefit. So the very act of storing and recalling information, of making and using a memory, is often entangled with the fact that the experience *matters*—which in conscious beings like us often means association with a feeling. Thus anhedonia could not only arise from the same process underlying dementia, but also impair memory itself, further increasing the correlation of these two states.

Many neuroscientists today believe that remembering involves a reactivation of some of the very same neurons that were active during the initial experience. Several investigators have used optogenetics to explore this idea, not in sensory regions of the brain but instead in memory-related structures called the hippocampus and the amygdala, by tagging cells that were strongly active during a learning experience (such as a fearful episode in a particular context), and then reactivating a subset of those tagged cells with light, much later, far from the fearful context across both space and time.

Mice can be seen exhibiting fear in these cases, even in the absence of anything related to the initial fear-inducing experience—that is, absent everything but the optogenetic reactivation of a few of the fear-memory neurons. So remembering, it seems, can happen when the right combination of brain cells—called an ensemble—speaks together.

If this is remembering, then what is the memory itself when it is not being actively remembered? Within which molecules, cells, or projections do the bits reside? Where is the actual information of a memory—of the stored experience, or knowledge, or feeling—as it lies dormant, awaiting recall?

Many in the field today think an answer to this question lies in a quantity called *synaptic strength*—a measure of how strongly a neuron can influence another neuron, defined as the *gain* from broadcaster to receiver. The stronger a synapse, or functional connection, between two cells, the greater the response in the receiving cell will be to a fixed pulse of activity in the broadcasting cell. As abstract as it seems, this change in influence at the synapses could *be* the memory, in a real and physical sense.

There are many interesting features to synaptic strength that make this idea plausible. First, theoretical neuroscientists have proven that synaptic strength changes indeed can store memories in an automatic way during experiences (without requiring intelligent supervision) and in a form that allows easy recall. Second, synaptic strength changes of the right kind can happen in the real world—in fact, very readily and swiftly in living neurons and brains—in response to bursts of activity or neurotransmitters. Certain patterns of synchronous or high-frequency activity pulses can drive increases—potentiation—in synaptic strength, while asynchronous or low-frequency pulses can drive decreases—depression—in synaptic strength. Both effects are plausibly useful for memory storage, based on the theoretical work.

It had only been a tantalizing hypothesis, that synaptic strength along a path from one part of the mammalian brain to another could be specifically and directly adjusted to change behavior. This idea had not been formally testable without a way to selectively provide activity pulses to change synaptic strength in projections defined by origin and target in the mammalian brain. But optogenetics enabled this intervention: a connection from one part of the brain to another can be made light-sensitive, and then high- or low-frequency light pulses can be provided along these pathways. By 2014 several groups working with mammals were optogenetically applying these principles from memory, and had confirmed that powerful and selective effects on behavior can be exerted by projection-specific synaptic strength changes themselves.

Projections fundamentally embody how effectively different parts of the brain can engage with each other, whether in health or disease; for example, it is known that interregional connectivity strength predicts interregional activity correlations. It is also known that interregional activity correlations can be linked to specific states of enjoyment—for example, reduced coordination between the auditory cortex and a deep reward-related structure (the nucleus accumbens) predicts anhedonia for music in human beings. Likewise, the specific basic reward of caring for a grandchild could be enabled by a capability for strong synaptic connectivity (and thus effective engagement) between one brain region responsible for addressing drives or rewards (like the hypothalamus or VTA/nucleus accumbens circuitry) and another brain region representing hierarchies of kinship relationships (like the lateral septum). Projection-specific synaptic strengths may allow such specific behaviors to become favored and rewarding, especially with learned positive experience.

In this way, synaptic strength at the level of brain region interconnection is an interesting quantity relevant to development and evolution of our internal feelings, since evolution is well suited to working with such interregional connection strengths. Though evolution knows nothing of music or grandchildren per se, it could set up the conditions allowing either or both to be enjoyed—to a certain level, with the right life experience. And there is no shortage of genetic complexity available to lay these specific foundations, in the richness of gene expression patterns that determine how cellular diversity and axon guidance implement brain wiring.

Value—whether negative for aversion, or positive for reward—in the end is only a neural label of sorts, one that can be attached to, or detached from, elements like experiences or memories. This flexibility is crucial for learning, for development, and for evolution. But what can be readily attached may be just as easily detached—for good or ill, in health or disease—and we now have a path for understanding how this flexibility can be enabled. Memories and values alike may both reside in synaptic strengths, learned or evolved as physical structures. And the path to the synapse—along the axon, the long-range fiber emerging from one cell to touch other cells—is set up and directed and grown according to instructions from genes (which follow all of evolution's rules) at which point the synapse itself can be powerfully tuned by the specificity of experience. Our paths, our joys, our values all lie along thin threads that can be cut—connections bearing our memories, projections that are our selves.

•

I signed out to the night psychiatry resident, whose Saturday swing shift was sandwiched between my two daytime Saturday and Sunday shifts. I had not met him before; he looked excessively sporty and energetic. Tired but thinking myself tolerant, I walked him through a summary of the patients on our unit with active issues before driving home to get a few hours of rest.

Driving back to the hospital again the following morning, along the deserted streets of early Sunday in Palo Alto, my thoughts drifted back to Mr. N. There remained challenging logistical issues if we were to initiate administration of a medication. We had to determine who was legally able to provide consent, and if Mr. N. was not able, the primary team needed to have a discussion with his son—someone I had not yet met. There was little I could do in the moment; I was technically only a consultant on this case, not a decision-maker.

After getting sign-out from the now-haggard night resident for the patients on our psychiatry unit and managing to listen with benign interest as he troweled on thickly the stories of his overnight valor, I went to a workstation to see if anything new had appeared regarding Mr. N. His location had changed, surprisingly—his name no longer appeared on the 4A medicine unit roster. A moment later I saw he was in the ICU—the intensive care unit.

Mr. N. had suffered a massive stroke the night before, an hour after I had left him. His body was alive, but it was unlikely he would ever recover independent living. His son had power of attorney. Code status had been set: do not resuscitate; do not intubate.

I stood dumb, staring, impotent. He had been right and needed to tell me. His night would be very long.

•

Only at the very end of life—only when we have put away the chess-board, with all moves made, with no more surprises to come, with most consequences played out—can we fairly judge ourselves and assign credit to actions that ultimately brought success or failure. But it is also here at the end where the memories of our moves fall away, forgotten—a cruel twist, for without memory, how then can we make sense of the lives we have lived, and find meaning in the paths we took, amid the pathos?

We cannot, and so we end where we begin, helpless and uncertain.

Mr. N. surprised us by living on for several weeks before passing. I saw a man whom I presumed to be his son two or three times around the hospice unit, walking in and out—and once as he was pushing Mr. N., supine on his gurney, down the corridor. I remember that day pausing to watch them, as they eased toward a patch of sunlight by a window, and I remember hearing his son's gentle whisper: *Here's some sun for you, Daddy.*

Mr. N. looked older than I remembered—lying flat, utterly limp, his skin a paler gray, eyes closed and mouth open, atonal, utterly still. Cashed out and gone home. His bristly head, the only part of him not covered with the blanket and sheet, seemed proud and dignified to me, though. It evoked the memory of his final move, sitting up in his hospital bed, telling me something that mattered through the fog of dementia and depth of depression, with nearly everything already taken away from him.

As they neared the window and its broad sunbeam, I could hear a medicine team hustling our way, chattering about atrial flutter. Mr.

N.'s son could hear them too—he pushed a little faster to make space, clumsily guiding the gurney toward the window at the corridor's edge.

As the team swept past me, humming along in a crescendo of discussion, the gurney jolted gently to a halt as its bumpered corner tapped the wall. At that moment of impact, both of Mr. N.'s arms suddenly swept up toward the ceiling, askew but together—the sheet dropping away as one arm ended firmly skyward, the other weaker, halfway up, both hands open and fingers splayed. Stable and strong. A frantic reach, a shocking strength.

A stunned moment of silence seized the hallway, and its motley collection of spectators, as Mr. N.'s son and the interns and I looked on at the reaching, grasping arms, all of us locked together in the surreal scene for a beat or two—and then the arms eased back to the gurney, together. Mr. N. was once again at rest.

The medicine team had slowed but never stopped. It took a few seconds for their chatter to build and re-form, as they turned a corner at the end of the hall, but now humming in a minor key, the neurology of reflexes surfacing in their minds from a whirlpool of memories and desires.

•

In dementia, the infantile reflexes come back, movements choreographed by evolution for the survival of primate babies: the *Moro* reflex (arms sweep up when the body is suddenly dropped or accelerated, a relic from our tree-climbing forebears, saving infant lives of those who became our ancestors) and the *root* (light touch on the cheek triggering a turn of the head and an opening of the mouth, to find milk). Falling from height, and losing contact with mother—the basic unlearned fears of human newborns.

Both action patterns disappear after a few months of life but come back with dementia or brain damage—not re-created at the end of life, but rather never really gone, always present but latent for decades, layered over with higher function, coated with inhibition and cognitive control, with all the threads of lived life. As the fabric frays and texture is lost, the original self finds voice again in a heart-wrenching grasp for safety, reaching for a long-dead mother.

All the details of life that mattered so much over the years, that brought moments of happiness or pain, had only covered her over, weaving across with so many weft threads that she could no longer be seen. But always there, and now at the end, the framework for everything surfaces again. As the fine threads fall away, she becomes, once more, the whole world. She might be reachable again, the mammal who kindled her baby to life, who shakingly held, and nursed, and shielded her child—from the rain and sun.

As the threads of the mind disintegrate, as massive insulated fibers fragment and fray, when memory and agency have dissolved away, what was there since birth is all that is left . . . a human infant in a thin cloth of gray, now exposed again to the cold.

Now all there is, in the confused darkness, is a gentle sway. . . . And when balance suddenly shifts, as the dry weak branch snaps, the baby is released into the night, unpinned from the world, and falls—hands sweeping, in a desperate grasp.

A branch breaks, and this is all at the end. A tree-dwelling baby, a grasping for mother, a falling through space.

EPILOGUE

My great blue bedroom, the air so quiet, scarce a cloud. In peace and
silence. I could have stayed up there for always only. It's something
fails us. First we feel. Then we fall. And let her rain now if she likes.
Gently or strongly as she likes. Anyway let her rain for my time is
come. . . .

So. Avelaval. My leaves have drifted from me. All. But one clings
still. I'll bear it on me.

—JAMES JOYCE, *Finnegans Wake*

The shuttle swings on, ticking back and forth at the leading edge of
the tapestry, marking time in space like a pendulum, embedding
moments and feelings. Warp threads point the way into unformed
space, framing—but not determining—what happens next.

This steady progression of experience clarifies patterns and buries
structural threads. Either outcome serves as a resolution of sorts.

My oldest son, with whom I lived as a single father for many of these
experiences—and whose broken home frightened me in the context of
what I was seeing clinically—has now grown into a hardworking computer scientist and medical student, with caring relationships and a
talent for guitar. Intersecting threads can either interrupt, or create, a
pattern—and life gives no explanation. I now have four lovely younger
children with an eminent physician-scientist—also at my university—
whose mission is studying, and treating, the very same brainstem tumor
that had grown in the little girl with eyes misaligned, who had almost
brought to an end my path in medicine. At the heart of every story here,
there is a lost child—but one who might still be found.

Every sensation described here, each individual feeling and thought
that guided me to this point, now seems more richly textured than
when first experienced, and more deeply interwoven. But is the original feeling better defined by, or instead obscured by, these connections

formed with time? In some ways it doesn't matter—any more than buried warp threads can be meaningfully revealed without destroying a tapestry, any more than we can expose and experience our original feelings again without cutting connections and memories, and losing ourselves.

Ongoing scientific developments will continue to provide more textured interpretations for the stories told here. With each new discovery, our own construction by evolution becomes ever less simply described, and even the extinction of Neanderthals acquires more dimensionality as paleogenetics progresses. Of course they live on in us, and so are not extinct in any definitive sense, but an even more profound truth has now become apparent. We now know that when the last Neanderthals died, they were already part modern human—because the intermingling went both ways, and the last Neanderthal may have been also the last survivor of a wave of modern humans that had left Africa first. Their extinction is truly human, our own.

Most of the medical discoveries described here will, over time, be identified as just elements of a much larger picture—and those will be the success stories. A few will be forgotten, or found flawed enough to require patching and replacement. But this process of discovery and repair of flaws in our understanding is identical with the progress of science. Gaps and flaws by their very nature—like the disease processes themselves—illuminate and reveal.

Light in the natural world passes only through gaps already present— like cracks in the cloud cover or passages through the forest canopy blown open. But with this biology, and in these stories, visible light inverts that paradigm by physically opening a gate—information creating a path for itself, illuminating the whole human family as it streams through. Sometimes it seems the channel is only clumsily stuck open, a rural cow gate in damp sod; we have not fully prepared the pathway, or ourselves, to deal with the information coming through. But the gate is open.

Recent years have even brought insight into the gate itself. Hearkening back to feelings experienced near the very beginning of my scientific journey—crossing scales, exploring mysteries of the whole brain while grounded all the way down at the cellular level in our scientific methods—we have also now delved even deeper, to the molecular and atomic levels of resolution, in exploring how the light-gated protein called channelrhodopsin actually works. We have been able to elucidate the mystery of how light can be detected by a molecule and then turned into electrical current flowing through a pore in that same individual molecule. These experiments use intense beams of X-rays—the same kind of scientific method, crystallography, that enabled the discovery of the double-helical structure of DNA.

There had been intense controversy: some prominent investigators had claimed that there was not a light-gated pore within the channelrhodopsin molecule. But X-ray crystallography allowed us not only to see the pore directly and prove its existence, but also to use that knowledge to redesign the pore and show the depth of our understanding in many ways: changing atoms around—reupholstering the pore's inner lining—to create channelrhodopsins that conduct negatively charged instead of positively charged ions, or to make these molecules responsive to red rather than just blue light, or to change the timescale of the electricity elicited, speeding or slowing the currents manyfold. These new channelrhodopsins have already proven useful for neuroscience across a broad range of applications, and so cracking the structural code of this mysterious light-gated channel—thereby solving a fundamental mystery rooted in the basic biology of a most amazing plant—also opened up a scientific pathway to new explorations of the natural world, and of ourselves.

Today, even as science in my own lab at Stanford progresses, I still treat patients in an outpatient clinic focusing on depression and autism (and serve as attending physician for a block of inpatient call time every year)—all the while working with a new generation of psychiatry residents, teaching and learning as we journey together through a field that still feels as enthralling and mysterious as it did in my first moment with the schizoaffective disorder patient. We achieve cures in many patients, and in others we can only manage symptoms—a path followed in many fields of medicine, where we manage intractable diseases because we can, and because if we don't, the patient dies. We are honest peddlers of herbs that help—of feverweed and foxglove.

As our understanding of psychiatry and our insight into the neural circuit control of behavior progress together, we might be wise to begin awkward conversations for which we feel unready. We will need to stay philosophically, and morally, ahead—rather than trying to catch up once it is too late. An uncertain world is already demanding from psychiatry answers to difficult questions about ourselves in health, not only in disease. The reasons for this pressure are important—to discover, and then grapple with, and then embrace, humanity's uplifting and disturbing contradictions.

And so here, in the form of epilogue, we can look briefly to the future, along three dark and deeply forested paths only dimly illuminated by the stories in this volume that each need more profound exploration soon: our process of science, our struggle with violence, and the understanding of our own awareness.

•

Scientific breakthroughs are difficult to predict or to control—forming an odd contrast with much of the process of science, which is an exercise in controlled, and ordered, thought. Indeed, ordered thought seems natural to the human mind in general, and control over the flow of complex thoughts is taken for granted, just as we assume the steady forward passage of time. And yet we cannot use our appetite for order and control to plan the process of science completely. This is a major lesson from most scientific breakthroughs, including that of optogenetics—revealing the need to support basic research that at some level is unplanned. It would have been impossible to predict the impact—on neuroscience—of research into microbial light responses over the past 150 years. Similarly unexpected developments have launched many scientific fields; indeed, because this volume is partly memoir the stories focus on optogenetics, but other pioneering fields have also converged from unexpected directions to define the landscape of biology today.

Thus optogenetics has revealed not only a great deal about the brain but also, in an accessible way, the nature of the basic scientific process. This idea is important to keep at the forefront of our minds as we move together into the future: scientific truth—a force that can rescue us from weaknesses of our own construction—arises from free expression and pure discovery. That, and perhaps also from a little bit of disordered thinking.

I recall an alcoholic cirrhosis patient, in my care but without prospect of a new liver, who is reaching his endgame. He is drowning on dry land in liquid of his own creation. His belly is turgid and tense with ascites—the brownish-yellow fluid of hepatic failure, perhaps ten liters or more, distending his abdomen, compressing his lungs and diaphragm from below. He is only forty-eight but struggling to breathe, gasping in the bed before me.

I hold a crude tool, a trocar. Medieval, heavy in my hands, it is all we have. Illuminated by the harsh, bright procedure light at the bedside, the trocar is still dull as pewter—sterile but stained, tarnished, a blunt cylinder for placing a drain in the abdominal wall. I can take off five or six liters of fluid from his abdomen at a time, but this buys him space to breathe for only two or three days before the steady accumulation of ascites fills him up again. I cannot cure the illness, but I can do something, steadily and carefully, until we know more.

Truth is our trocar, for now. Truth that we can get to, as soon as we can get to it, through the open conversations that we know, of free argument and creative discovery.

Science, like songs and stories, serves as such a free form of human communication. Science also differs in that the conversation seems at first limited to the fraction of human beings trained to appreciate its full meaning. But like the performance artist Joan Jonas said about her art in 2018, science is "a conversation with the past and the future, and with a public." Scientists are not recluses shouting data into the void, nor automatons filling drives with bits. We seek truth, but truth to communicate in ways that we think and hope may matter. The meaning in our work comes from human partners we imagine and direct our voice toward, with awareness that these conversations will not be one-way.

Even completing a major breakthrough requires understanding of how it will be communicated, which in turn requires consideration of the listeners as well as the speaker, and the volatile context—the dynamic frame of the world beyond, its time and place in the human story. In unformed open spaces that bring no judgment or posturing, our path forward is that of a patient in talk therapy, where insight is achieved only through engaging freely and frankly, and without possibility of penalty. Otherwise immature defenses are the resort; walls are erected, evincing no understanding, isolating our own feelings—walls built because honest and free conversation, engaging all human family members, has not been prioritized. We need to be what we might be, so we can discover who we are.

•

Not far beneath the surface, part of what we can be is violent toward one another. There are many—too many—paths that lead to violence, with societal complexity that is important to understand, and this is perhaps the subject of another, different text. But when violence is visited upon human beings by human beings, without an obvious reason, seemingly for its own sake, then psychiatry (and thus neuroscience) would seem as close to the front lines as any school of human thought. This situation is usually framed in psychiatry as antisocial personality disorder, which is largely overlapping in meaning with *sociopathy*, a term widely used in the general public. We have no answers to the natural human questions—why does this disorder exist, and what can be done?—while it seems the need to understand this condition grows more urgent every day.

What proportion of humanity is capable of causing pain, or death, with utter disregard for human feeling? Estimates vary enormously depending on the study or population, from 1 to 7 percent, likely due to matters of degree—and variance in opportunity, which may be all that separates the active cases from those in a latent state.

In psychiatry, the definition of antisocial personality disorder includes "a long-term pattern of disregard for, or violation of, the rights of others," and so the criteria can be met in a person with both cruelty to animals as a child and disregard for physical or psychological integrity of other human beings as an adult. Both bits of history can be concealed, but both often are uncovered with surprising ease in the psychiatric interview, and the trained psychiatrist can come to a tentative diagnosis quite quickly.

What do we do with 1 to 7 percent: this high number, and this broad range? Are we good at heart, or original sinners? Either way, a strong argument makes itself for setting up societies so that no one person is ever fully trusted or empowered—with checks at all levels: personal, institutional, and governmental. But at even a few percent, this means the condition is deeply baked into the population. This seems a heavy burden for our species, explaining a great deal about human history and the present day—but one hopes, less so about the future, for how else can one imagine a future for humanity?—as the consequences of our actions become more global and more permanent.

Astrophysicists ask a related question in thinking about the cosmos: with its innumerable planets and its billions of years, if total technological transformation of a species and of a world takes only a few hundred years as we know, just a moment, an eyeblink, why does the universe seem so quiet? An easy explanation is that extinction follows very quickly from technology. No amount of institutional restraint is ever enough—the drives that support survival, in the end also drive extinction. Evolution creates intelligence that is unsuited for the world that intelligence, in turn, creates.

Can deeper scientific understanding of the biology save us? Little is known about the biology of antisocial personality. There is a heritable component (accounting for as much as 50 percent) revealed by studies of twins, and some evidence for reduced volume of cells in the prefrontal cortex, where aspects of restraint and sociability operate. Specific genes have been linked to sociopathy or aggression, including those encoding proteins that process neurotransmitters like serotonin in the synapse—and altered brain activity patterns have been observed, including changes in coordination between the prefrontal cortex and reward-related structures like the nucleus accumbens. But we lack deep understanding or clearly identified paths of action. Contradictions in the field still abound—for example, disagreement as to whether impulsive violence or its polar opposite (calculating, manipulative violence) is the more relevant core symptom. Each concept points to opposing ideas for diagnosis and treatment.

Modern neuroscience, however, has begun to illuminate the circuits underlying violence directed to another member of the same species—in studies that (while revealing) verge on the deeply disturbing. In a striking example of such a discovery that could not have been accomplished with prior methods, one group of researchers tried in rodents to test electrical stimulation of a tiny sliver of mammalian brain thought to modulate aggression—the ventrolateral subdivision of the ventromedial hypothalamus, or VMHvl. The research team was unable to observe aggressive responses despite numerous attempts at stimulating with an electrode, likely because the VMHvl is a small structure that is closely surrounded by other structures that elicit defensive measures instead, such as freezing or fleeing; these surrounding structures or their fibers would also be activated by electrical stimulation of the VMHvl, confounding and confusing the behavioral results. But when the team next made use of the precision of optogenetics to target only VMHvl cells with an excitatory microbial opsin, stimulation of these cells with light elicited a frenzy of violent aggression toward another mouse in the cage (a smaller nonthreatening member of the same species, the same strain, that the optogenetically controlled mouse had been perfectly content to leave alone, until the moment of flipping on the laser light).

The fact that individuals can be so instantaneously and powerfully altered in their expression of violence points to deep questions of moral philosophy. In teaching optogenetics to undergraduates, I have found it striking to see the responses they exhibit upon seeing videos— peer-reviewed and published in major journals—of instantaneous optogenetic control of violent aggression in mice. Afterward, the students often need a period of discussion, almost a dose of therapy, simply to process and incorporate into their worldviews what has been observed.

What does it mean about us, that violent aggression can be so specifically and powerfully induced by turning on a few cells deep in the brain? As the professor I can transmit the perspective that this is not entirely a new effect—aggression has previously been modulated to varying degrees, over decades, with genetic, pharmacological, surgical, and electrical means. But such knowledge seems to be of little value to the students in the moment. With these prior interventions, there has always been a veneer of nonspecificity and side effects. In contrast, the more precise an optogenetic intervention becomes, in the context of a seeming lack of self-limitation, the more problematic are its implications, and the better posed certain conundrums seem to be.

And what exactly are those conundrums? Optogenetics is too complex to be a weapon; rather, the issue is what the animals seem to be telling us about ourselves: the change in violent behavior, in its power and speed and specificity, seems disconnected from, or unconnectable to, ways we seek to combat violence in our civilization—that is, these powerful neural circuit processes seem destined to ultimately overpower fragile societal structures set up to prevent moral detachment. What can be done? What hope is there? What are we, really, when murderous violence can be instantly induced by only a few electrical blips in a few cells?

But violence can be suppressed by a few spikes too, and so at least there is now a path forward: using optogenetics and related methods to elucidate the cells and circuits that *suppress* aggression. And even if not immediately practical or therapeutic, this neuroscience-based dimension of insight enables us to move beyond intense societal debates of the past (while building on what has come before). We can now begin to unify the intersecting influences of genes and culture in a concrete and causal framework. We now understand enough about behavioral causation to see how elements of neurophysiology underlying behaviors as complex as violent aggression can be manifested in well-defined physical components of the brain: projections endowed with form (direction and strength) by individual brain development on the one hand, and learned life experience on the other.

Because we do not fully control either our brain development or our life experience, the precise nature of personal responsibility for action is still an interesting and contentious matter. A modern neuroscience perspective informed by the kind of work described in this volume might hold that there is no personal responsibility for some actions that involve the brain (like startle responses) because circuitry involving the self was never consulted, in contrast to other kinds of action where priorities and memories weigh in—that is, where circuits are engaged that define one's path through the world, such as the retrosplenial cortex and prefrontal cortex. Since such a sentence, describing causal and measurable concepts, can be reasonably written without using words like *consciousness* or *free will*, which are difficult to quantify, modern neuroscience may indeed be able to make headway on these hard questions, that until now have inhabited only the fascinating domain of the philosophical treatise.

There is unlikely to be a single location in the brain that explains freely chosen actions; indeed, we are increasingly able to grapple with more widely distributed circuitry for decision-making and path-selection, as we attain ever-broader perspectives on the activity of cells and projections throughout the brain during behavior. In 2020, recording the activity of cells broadly across mouse and human brains brought insight into the circuit-level construction of the self, by probing the fascinating process of dissociation—in which the inner sense of one's self is separated from physical experience, and so individuals feel dissociated from their own bodies. The self is aware of but detached from sensations—no longer feeling ownership of, or responsibility for, the body. With optogenetics and other methods, activity patterns in the retrosplenial cortex (consistent with ideas described in the story on eating disorders) and certain of its far-flung projection partners were found to be important for regulating the unified nature of the self and its experience. And so one can accept that there may be a distributed origin of any action, and of the self as well, without abandoning the idea of the self as a real and biological agent subject to precise scientific investigation.

Meeting this complexity head-on may eventually allow us to understand and treat (and feel empathy for) the antisocial, who may have as much free will and personal responsibility as any other person but who can often be cruel to themselves, to their own selves, as well—perhaps through a biologically definable form of detachment, or dissociation, from the feelings of both self and others. As a physician, understanding this last trait—more than any other aspect—helps me care as I should, and do, for these fellow human beings, despite everything.

·

The future of this scientific journey—given our accelerating progress toward accessing all the cells, connections, and activity patterns of brain cells in animals during behavior—is leading us not only to understanding and treatment of our own difficult and dangerous design, but also to insight into one of the most profound mysteries of the universe. Rivaling the question of *Why are we here?* is *Why are we aware?*

In 2019 optogenetic technology began to enable control of mammalian behavior in a wholly new way, no longer just allowing control of cells by type—the workhorse of optogenetics for its first fifteen years—but also allowing control of activity in many single cells, or individually specified neurons. Now we can pick, at will, tens or hundreds of single cells for optogenetic control—cells selected from among millions of neighbors by virtue of their location, their type, and even their natural activity during experiences.

This effect was achieved by developing new microscopes, including holographic devices based on liquid crystals. These machines take a massive leap beyond fiberoptics to use holograms as the interface between light and the brain, allowing a sort of sculpting of complex distributions of light even in three dimensions—to control individual opsin-producing neurons during behavior in a mammal such as a mouse.

In one application of this method, we can cause animals in complete darkness to behave as if they were seeing specific visual objects of our own design. For example, we can pick the cells that normally respond to vertical (but not horizontal) stripes in the visual environment and then, without any such visual stimulus, optogenetically turn on just those cells with our holographic spots of light, testing to see if the mouse acts as if there were vertical stripes present. Both the mouse and the mouse's brain in fact behave as if the vertical stripes were there; peering into the activity of many thousands of individual neurons in the primary visual cortex (that part of the cortex that first receives information from the retina), we can see that the rest of this circuitry—with all the complexity of its immense number of cells—acts as it does during natural perception of real vertical (but not horizontal) stripes.

We now find ourselves in an astonishing position: we can pick out groups of cells that are naturally active during an experience and then (using light and single-cell optogenetics) insert their activity patterns back in *without* the experience; when we do so, the animal (and its brain) both behave naturalistically as they would when perceiving a real stimulus. The animal behavior showing correct discrimination is similar whether the sensory stimulus is natural or provided entirely by optogenetics—and the detailed, real-time, cellular-level internal representation of the sensory discrimination across volumes of the brain is also similar, whether the sensory stimulus is natural or provided entirely by optogenetics. As far as we can tell, then (with the caveat that I will never know what another animal, human or not, is truly experiencing subjectively), we are directly inserting something resembling a specific sensation as defined by natural behavior and natural internal representation.

We were intrigued to see how few cells we could stimulate to mimic the percept, and we found that a handful was enough—as little as two to twenty cells, depending on how well trained the animal had been. So few cells sufficed, in fact, that a new question had to be asked: why are mammals not frequently distracted by chance synchrony events among a few cells that happen to have similar natural responsivity— thus fooling the brain into concluding (wrongly) that the object these cells are designed to detect must be present? In some people that may happen, as in Charles Bonnet syndrome, where people who suffer adult-onset blindness can experience complex visual hallucinations; the visual system seems to act as if all is too quiet, reaching to create something, anything, out of the noise. I treated a patient at the VA hospital with this syndrome: an amiable and elderly veteran, entirely blind, who would see fully formed visions, often of sheep and goats grazing harmlessly in the middle distance. We found his visions could be reduced with an antiepileptic medication called valproic acid, but in the end we discharged him with no prescription; he had grown attached to what his deprived visual cortex had decided to provide for him.

More broadly, such spontaneous unwanted output—of any part of the brain due to spurious correlations of a few of its cells—could be a principle underlying many psychiatric disorders, ranging from the actual auditory hallucinations of schizophrenia, to unwanted motor outputs and thoughts of tic disorders and Tourette's syndrome, to out-of-control cognitions like those of eating and anxiety disorders. The mammalian brain is poised perilously close to the level where noise might escape to be treated like a signal—an insight important both for the basic neuroscience of natural mammalian behavioral variability and for clinical psychiatry.

Beyond science and medicine, philosophical puzzles surrounding subjective awareness suddenly become better posed with this multiple-single-cell control. Indeed, new life has even been breathed into philosophical thought experiments (*Gedankenexperimente*, as the physicists Ernst Mach and Albert Einstein might have said, following a tradition extending back at least to Galileo), ancient in formulation and discussion. The modernized version of an old story might run as follows:

Suppose one could control (as with this new form of single-cell optogenetics) the exact pattern of activity, over some period of time, of every cell in the brain of an animal capable of subjective sensation—say, of a pleasurable, intensely rewarding, internal feeling. And suppose such control could even be precisely guided by first observing and recording those activity patterns in the same animal during natural exposure to a real, rewarding stimulus—just as we already know we can do for simple visual percepts in the visual cortex.

The seemingly trivial question, then, is: would the animal feel the same subjective sensation? We already know that a mouse and its visual cortex will both behave as if it had received and processed the real stimulus—but would the animal also feel the same internal awareness, experiencing its quality beyond the information itself, like natural subjective consciousness except now when the activity pattern is presented artificially?

It is important that this be a thought experiment—of course we cannot fully know the subjective experience of another individual, even of another human being, nor have we yet achieved the total control contemplated here—but like Einstein's original *Gedankenexperimente* that illuminated relativity so powerfully, this thought experiment rapidly brings us to a conceptual crisis—one that, in its eventual resolution, could be highly informative.

The problem is that answering either yes or no to this question seems essentially impossible. Saying no implies there is more to subjective sensations than the cellular patterns of activity in the brain—since in the thought experiment, we are allowed to match the precise patterns of all physical phenomena that cellular activity elicits, including neuromodulators, biochemical events, and so on that are the natural consequences of neural activity. As a result, we have no framework to understand how that answer could be no. How can there be more to what the cells of the brain do than what they do?

Saying yes raises equally unsettling questions. If all the cells are actively controlled and a subjective sensation is being felt, then there is no reason the cells all need to be in the head of the animal. They could be spread all around the world, and controlled in the same way with the same relative timing, over as long a time period as was interesting, and the subjective sensation should still somehow be felt, somewhere, somehow, by the animal—an animal no longer existing in any discrete physical form. In a natural brain, neurons are near one another, or connected to one another, only to influence one another. But in this thought experiment, neurons no longer need to influence one another—the exact effect of what that influence would have been, over any period of time, is already being provided by the artificial stimulus.

This answer intuitively seems wrong, as well, though we are not sure how—it just seems to fail an absurdity test. How, and why, could individual neurons spread around the world still give rise to the inner feelings of a mouse or a human? The question is only interesting because we are considering an inner feeling. If instead we were to divide a basketball into a hundred billion cell-like parts to be distributed around the world and controlled individually to move as they would during a bounce, there would be no philosophical debate as to whether this new system feels as though it were bouncing. The answer would be, presumably: no more or less than the original ball.

We are left with a philosophical problem, one that optogenetics has framed sharply and clearly. There are certain to be many such mysteries about the brain, like the nature of our inner subjective states, that don't fall into current scientific frameworks: questions that are deep and unanswered, but some that—it seems now—can be well posed.

And those subjective states, called qualia or feelings, are not just abstract or academic concepts. These are the same inner states that were the central focus of this volume, that first brought me to psychiatry years ago, each one inseparable from its own projection across time—over seconds, and over generations. These subjective experiences underlie our common identity and define the path we have traveled, together, as humanity—even if shared only as stories, in a book or around a fire.

ACKNOWLEDGMENTS

I am deeply indebted to so many people who helped nurture this work, and who provided motivation and energy through difficult times.

Heartfelt gratitude to Aaron Andalman, Sarah Caddick, Patricia Churchland, Louise Deisseroth, Scott Delp, Lief Fenno, Lindsay Halladay, Alizeh Iqbal, Karina Keus, Tina Kim, Anatol Kreitzer, Chris Kroeger, Rob Malenka, Michelle Monje, Laura Roberts, Neil Shubin, Vikaas Sohal, Kay Tye, Xiao Wang, and Moriel Zelikowsky for their notes and comments—along with my perceptive and tireless literary agent Jeff Silberman, and my deeply thoughtful editor and publisher Andy Ward, whose belief in these stories was always greater than my own.

I am most grateful to all the people who shared this path with me— merging their stories with my own, for a time.

NOTES

Brief references, for background on the science within each story, are included here. All of the articles are freely accessible; you can either copy and paste the link into a browser search bar (if you're reading on a connected device) or for the notes labeled PMC (for PubMedCentral) go to https://www.ncbi.nlm.nih.gov/pmc/articles/ and at the search bar enter the digital identifier shown (for PMC4790845, enter 4790845), whereupon articles can be read online or free pdfs downloaded.

PROLOGUE

8 **actual memory storage needing no guidance or supervision:** https://en.wikipedia.org/wiki/Hopfield_network; https://en.wikipedia.org/wiki/Backpropagation.

12 **In optogenetics we borrow genes from diverse microbes:** https://www.ncbi.nlm.nih.gov/pmc/articles/PMC4790845/.

13 **tricks from chemistry are used to build transparent hydrogels:** https://www.ncbi.nlm.nih.gov/pmc/articles/PMC5846712/.

13 **All the interesting parts remain locked in place, still within 3-D tissue:** https://www.ncbi.nlm.nih.gov/pmc/articles/PMC6359929/.

13 **national and global initiatives to understand brain circuitry:** https://braininitiative.nih.gov/sites/default/files/pdfs/brain2025_508c.pdf; https://braininitiative.nih.gov/strategic-planning/acd-working-group/brain-research-through-advancing-innovative-neurotechnologies.

14 **many thousands of insights into how cells give rise to brain function and behavior:** https://www.ncbi.nlm.nih.gov/pmc/articles/PMC4069282/; https://www.ncbi.nlm.nih.gov/pmc/articles/PMC4790845/.

14 **connections defined by their origin and trajectory through the brain could now be precisely controlled:** https://www.ncbi.nlm.nih.gov/pmc/articles/PMC4780260/; https://www.ncbi.nlm.nih.gov/pmc/articles/PMC5729206/.

CHAPTER 1: STOREHOUSE OF TEARS

25 **they traveled into, and dwelt within, our cellular forebears more than two billion years ago:** https://www.ncbi.nlm.nih.gov/pmc/articles/PMC5426843/.

26 **with optogenetics, microbial DNA has yet again returned to animal cells:** https://www.ncbi.nlm.nih.gov/pmc/articles/PMC5723383/.

29 how much of modern Eurasian human genomes arose from this interaction—about 2 percent: https://www.ncbi.nlm.nih.gov/pmc/articles/PMC5100745/.

30 a hidden cave, alone in a final redoubt near the coast of Iberia: https://en.wikipedia.org/wiki/Gorham%27s_Cave; https://www.ncbi.nlm.nih.gov/pmc/articles/PMC6485383/; https://www.ncbi.nlm.nih.gov/pmc/articles/PMC5935692/.

31 from an extension of the amygdala called the bed nucleus of the stria terminalis: https://www.ncbi.nlm.nih.gov/pmc/articles/PMC6690364/.

32 A fiberoptic can be placed not in the BNST but in an outlying region: https://www.ncbi.nlm.nih.gov/pmc/articles/PMC4069282/; https://www.ncbi.nlm.nih.gov/pmc/articles/PMC3154022/; https://www.ncbi.nlm.nih.gov/pmc/articles/PMC3775282/.

34 In the mouse version of place preference: https://www.ncbi.nlm.nih.gov/pmc/articles/PMC5262197/; https://www.ncbi.nlm.nih.gov/pmc/articles/PMC4743797/.

35 Thus a complex inner state can be deconstructed into independent features: https://www.ncbi.nlm.nih.gov/pmc/articles/PMC6690364/.

35 parenting, in the form of intimate care of mammals for their young, was soon deconstructed into component parts, mapped onto projections across the brain: https://www.ncbi.nlm.nih.gov/pmc/articles/PMC5908752/.

37 Tears are powerful for driving emotional connection: https://www.ncbi.nlm.nih.gov/pmc/articles/PMC4882350/; https://www.ncbi.nlm.nih.gov/pmc/articles/PMC5363367/.

38 missing even this one part of the conversation may come at a cost: https://www.ncbi.nlm.nih.gov/pmc/articles/PMC4934120/; https://www.ncbi.nlm.nih.gov/pmc/articles/PMC6402489/.

39 collections of cells, sixth and seventh and parabrachial, jostle together in a small spot in the pons: https://en.wikipedia.org/wiki/Cranial_nerves.

43 In 2019 cells were studied across the entire brain of the tiny zebrafish: https://www.ncbi.nlm.nih.gov/pmc/articles/PMC6726130/.

43 optogenetics and other methods had implicated these same two structures in mammals: https://www.ncbi.nlm.nih.gov/pmc/articles/PMC5929119/.

43 Even the tiny nematode worm *Caenorhabditis elegans* appears to calculate: https://www.ncbi.nlm.nih.gov/pmc/articles/PMC3942133/.

46 Each mammalian species, on average, gets a run of about a million years: https://en.wikipedia.org/wiki/Background_extinction_rate.

46 population sizes around the world may have plummeted to a few thousand individuals: https://www.ncbi.nlm.nih.gov/pmc/articles/PMC5161557/; https://www.ncbi.nlm.nih.gov/pmc/articles/PMC4381518/.

CHAPTER 2: FIRST BREAK

49 the 767, slowly banking harborward, nearing the burning steel tower: https://en.wikipedia.org/wiki/United_Airlines_Flight_175.

55 Mood elevation has the capability to bring forth energy: https://www.ncbi .nlm.nih.gov/pmc/articles/PMC3137243/; https://www.ncbi.nlm.nih.gov/pmc/ articles/PMC2847485/.

57 mania is often not threat-triggered at all, and does not even approach utility: https://www.ncbi.nlm.nih.gov/pmc/articles/PMC2796427/.

57 *bouffée délirante* in West Africa and Haiti, a state of sudden agitated behavior: https://www.ncbi.nlm.nih.gov/pmc/articles/PMC4421900/.

59 broken fragments of the yolk genes persist, even within our own genomes: https://www.ncbi.nlm.nih.gov/pmc/articles/PMC2267819/; https://www.ncbi .nlm.nih.gov/pmc/articles/PMC5474779/.

59 Cave fish and cave salamanders—in sunless colonies, blocked off from the surface world: https://www.ncbi.nlm.nih.gov/pmc/articles/PMC5182419/.

60 dopamine neurons have attracted attention for their known roles in guiding motivation and reward seeking: https://www.ncbi.nlm.nih.gov/pmc/ articles/PMC4160519/; https://www.ncbi.nlm.nih.gov/pmc/articles/ PMC4188722/.

61 In 2015 the dopamine and circadian aspects were brought together with optogenetics: https://www.ncbi.nlm.nih.gov/pmc/articles/PMC4492925/.

61 the dopamine neuron population is not monolithic but composed of many distinct types that can be separably identified early in mammalian brain development: https://www.ncbi.nlm.nih.gov/pmc/articles/ PMC6362095/.

61 ankyrin 3 (also known as ankyrin G), which organizes the electrical infrastructure: https://www.ncbi.nlm.nih.gov/pmc/articles/PMC3856665/; https:// www.ncbi.nlm.nih.gov/pmc/articles/PMC2703780/.

62 In 2017 a mouse line was created with "knocked out"—insufficient— ankyrin 3: https://www.ncbi.nlm.nih.gov/pmc/articles/PMC5625892/.

CHAPTER 3: CARRYING CAPACITY

67 Two or three hundred milliseconds elapse before the response to a ping: https://www.ncbi.nlm.nih.gov/pmc/articles/PMC166261/; https://www.ncbi .nlm.nih.gov/pmc/articles/PMC4467230/.

74 Despite anxiety and cognitive impairment, Williams patients can seem extremely socially adept: https://www.ncbi.nlm.nih.gov/pmc/articles/ PMC4896837/; https://www.ncbi.nlm.nih.gov/pmc/articles/PMC3378107/.

80 "That incredibly thin, hairlike wasp waist": https://www.ncbi.nlm.nih.gov/ pmc/articles/PMC3016887/.

80 "ants and hornets and bees, all the social groups—later reverted away from

this life cycle": https://www.sciencedirect.com/science/article/pii/
S0960982217300593?via%3Dihub; https://www.sciencedirect.com/science/
article/pii/S0960982213010567?via%3Dihub.

81 researchers studying parenting in mice had used optogenetics: https://www
.ncbi.nlm.nih.gov/pmc/articles/PMC5908752/.

84 many of the genes linked to autism are related to these processes of electrical and chemical excitability: https://www.ncbi.nlm.nih.gov/pmc/articles/
PMC4402723/; https://www.ncbi.nlm.nih.gov/pmc/articles/PMC4624267/;
https://www.biorxiv.org/content/10.1101/484113v3.

84 people on the autism spectrum exhibit signs of increased excitability:
https://www.ncbi.nlm.nih.gov/pmc/articles/PMC4105225/.

85 speculated that a unifying theme in autism could be an increased power of
neuronal excitation—relative to countervailing influences like inhibition:
https://www.ncbi.nlm.nih.gov/pmc/articles/PMC6748642/; https://www.ncbi
.nlm.nih.gov/pmc/articles/PMC6742424/.

86 elevating the activity of *excitatory* cells in the prefrontal cortex caused an
enormous deficit in social behavior: https://www.ncbi.nlm.nih.gov/pmc/
articles/PMC4155501/.

86 mice (altered in a single gene called *Cntnap2*): https://www.ncbi.nlm.nih
.gov/pmc/articles/PMC3390029/.

87 this autism-related social deficit could be *corrected* by optogenetic interventions: https://www.ncbi.nlm.nih.gov/pmc/articles/PMC5723386/.

89 causing high excitability of prefrontal excitatory cells (an intervention that
elicited social deficits) actually did reduce the information-carrying capacity of the cells themselves: https://www.ncbi.nlm.nih.gov/pmc/articles/
PMC4155501/.

91 "The tree thrives in salt, and makes the soil salty too": https://www.ncbi
.nlm.nih.gov/pmc/articles/PMC5570027/; https://www.ncbi.nlm.nih.gov/pmc/
articles/PMC4836421/.

94 PTSD (a common and deadly disease that is often resistant to treatment
by medication): https://www.ncbi.nlm.nih.gov/pmc/articles/PMC5126802/.

CHAPTER 4: BROKEN SKIN

107 Skin arises from ectoderm, in embryos: https://en.wikipedia.org/wiki/Germ
_layer.

107 a meteor strike upended everything: https://www.youtube.com/watch?v=
tRPu5u_Pizk.

108 sensory skin organs then spread across the body: https://www.ncbi.nlm.nih
.gov/pmc/articles/PMC4245816/.

109 the patient or the psychiatrist might fit into a role from the past: https://
www.ncbi.nlm.nih.gov/pmc/articles/PMC6481907/.

110 Suicide is more common in borderline than in any other psychiatric disorder: https://www.ncbi.nlm.nih.gov/pmc/articles/PMC4102288/.

110 an unjust burden: psychological or physical trauma at a young age: https://www.ncbi.nlm.nih.gov/pmc/articles/PMC3402130/.

111 trauma during dependency—early in life when warmth and care are needed at all costs—predicts nonsuicidal self-injury later: https://www.ncbi.nlm.nih.gov/pmc/articles/PMC5201161/.

111 Our brains are building even basic structure—the electrical insulation, the myelin: https://www.sciencedirect.com/science/article/pii/S0092867414012987?via%3Dihub.

112 an individual can instead be guided chiefly by suppression of internal discomfort as the motivation for action: https://www.ncbi.nlm.nih.gov/pmc/articles/PMC5723384/.

113 cause animals to become more or less aggressive, defensive, social, sexual, hungry, thirsty, sleepy, or energetic: https://www.ncbi.nlm.nih.gov/pmc/articles/PMC5708544/; https://www.ncbi.nlm.nih.gov/pmc/articles/PMC4790845/.

113 swift to react strongly with value assignments: https://www.ncbi.nlm.nih.gov/pmc/articles/PMC5472065/.

114 causes the animal to begin to avoid the harmless room, as if it were a source of intense suffering: https://www.ncbi.nlm.nih.gov/pmc/articles/PMC4743797/.

114 turning down the dopamine neurons in the midbrain naturally, just as optogenetics does experimentally: https://www.ncbi.nlm.nih.gov/pmc/articles/PMC3493743/.

115 early-life stress and helplessness can increase habenula activity: https://www.ncbi.nlm.nih.gov/pmc/articles/PMC6726130/.

120 outpatient referral for a specialized group behavioral therapy: https://www.ncbi.nlm.nih.gov/pmc/articles/PMC6584278/.

CHAPTER 5: THE FARADAY CAGE

123 installing a true modern Faraday cage as a shield: https://en.wikipedia.org/wiki/Faraday_cage.

135 the Kalman filter, an algorithm for modeling complex unknown systems: https://en.wikipedia.org/wiki/Kalman_filter.

137 "Optimal filters will still block a few things that you actually wanted to go through": https://en.wikipedia.org/wiki/Chebyshev_filter; https://en.wikipedia.org/wiki/Butterworth_filter.

141 Matthews had imagined something he called an "Air Loom": https://en.wikipedia.org/wiki/James_Tilly_Matthews.

147 schizophrenia genetics: the collection of DNA sequence information from

human genomes: https://www.ncbi.nlm.nih.gov/pmc/articles/PMC4112379/; https://www.ncbi.nlm.nih.gov/pmc/articles/PMC4912829/.

150 evidence that disease symptoms are more common and strong in city dwellers: https://www.ncbi.nlm.nih.gov/pmc/articles/PMC3494055/.

CHAPTER 6: CONSUMMATION

165 cognitive and behavioral therapies can help in anorexia nervosa: https://www.ncbi.nlm.nih.gov/pmc/articles/PMC6181276/.

165 Medications are used not as cures, not to strike at the heart of the disease: https://www.ncbi.nlm.nih.gov/pmc/articles/PMC4418625/.

165 then eating disorders together show the highest death rates of any psychiatric disease: https://www.ncbi.nlm.nih.gov/pmc/articles/PMC2907776/.

168 the diversity of genes that can be involved, as with many psychiatric disorders: https://www.ncbi.nlm.nih.gov/pmc/articles/PMC5581217/; https://www.ncbi.nlm.nih.gov/pmc/articles/PMC6097237/.

168 Controlling the walking rhythms of the brainstem and spinal cord: https://www.ncbi.nlm.nih.gov/pmc/articles/PMC5937258/; https://www.ncbi.nlm.nih.gov/pmc/articles/PMC4844028/.

177 the first optogenetic control of free mammalian behavior was in the hypothalamus: https://www.ncbi.nlm.nih.gov/pmc/articles/PMC6744371/.

178 cause the hunger or thirst behaviors, actually driving the consumption of food or water: https://www.ncbi.nlm.nih.gov/pmc/articles/PMC5723384/.

180 specific social cells could suppress feeding, driving a resistance, even in naturally hungry mice: https://www.ncbi.nlm.nih.gov/pmc/articles/PMC6447429/.

180 When a mouse is fully water-sated but the deep thirst neurons are optogenetically driven: https://www.ncbi.nlm.nih.gov/pmc/articles/PMC6711472.

182 entorhinal cortex and hippocampus, two structures involved in navigation and memory: https://escholarship.org/uc/item/4w36z6rj.

182 a human subject is asked to sit quietly and think of nothing in particular, to simply be with one's self: https://www.ncbi.nlm.nih.gov/pmc/articles/PMC1157105/.

CHAPTER 7: MORO

188 infarcts, spots of dead tissue that are the outcome of strokes, can be detected by computed tomography: https://en.wikipedia.org/wiki/Vascular_dementia.

188 magnetic resonance imaging (MRI) can show the small vessel blockade of vascular dementia: https://www.ncbi.nlm.nih.gov/pmc/articles/PMC3405254/.

194 **350 million years ago, when the first air-breathing fishes emerged onto land:** https://www.ncbi.nlm.nih.gov/pmc/articles/PMC3903263/.

195 **global collapse in human populations bottoming out only fifty thousand years ago:** https://www.ncbi.nlm.nih.gov/pmc/articles/PMC5161557/; https://www.ncbi.nlm.nih.gov/pmc/articles/PMC4381518/.

196 **the few medications available only slightly slow the steady progression of the disease:** https://www.ncbi.nlm.nih.gov/pmc/articles/PMC6309083/.

197 **anhedonia in senior populations with cognitive impairment:** https://www.ncbi.nlm.nih.gov/pmc/articles/PMC2575050; https://www.ncbi.nlm.nih.gov/pmc/articles/PMC4326597/.

197 **the greater the accumulated volume of those lacunae in white matter:** https://www.ncbi.nlm.nih.gov/pmc/articles/PMC2575050/.

198 **valence of release from anxiety is set in part by projections from the BNST to reward circuitry:** https://www.ncbi.nlm.nih.gov/pmc/articles/PMC6690364/.

198 **absent everything but the optogenetic reactivation of a few of the fear-memory neurons:** https://www.ncbi.nlm.nih.gov/pmc/articles/PMC3331914/; https://www.ncbi.nlm.nih.gov/pmc/articles/PMC6737336/; https://www.ncbi.nlm.nih.gov/pmc/articles/PMC4825678/.

199 **synaptic strength changes indeed can store memories in an automatic way:** https://en.wikipedia.org/wiki/Hopfield_network; https://en.wikipedia.org/wiki/Backpropagation.

199 **synaptic strength changes of the right kind can happen in the real world:** https://www.ncbi.nlm.nih.gov/pmc/articles/PMC1693150/; https://www.sciencedirect.com/science/article/pii/S0092867400804845?via%3Dihub; https://www.ncbi.nlm.nih.gov/pmc/articles/PMC1693149/.

199 **Both effects are plausibly useful for memory storage, based on the theoretical work:** https://www.ncbi.nlm.nih.gov/pmc/articles/PMC5318375/.

199 **a connection from one part of the brain to another can be made light-sensitive, and then high- or low-frequency light pulses can be provided:** https://www.ncbi.nlm.nih.gov/pmc/articles/PMC3154022/; https://www.ncbi.nlm.nih.gov/pmc/articles/PMC3775282/; https://www.ncbi.nlm.nih.gov/pmc/articles/PMC6744370/.

200 **selective effects on behavior can be exerted by projection-specific synaptic strength changes:** https://archive-ouverte.unige.ch/unige:38251; https://archive-ouverte.unige.ch/unige:26937; https://www.ncbi.nlm.nih.gov/pmc/articles/PMC4210354/.

200 **Projections fundamentally embody how effectively different parts of the brain can engage with each other, whether in health or disease:** https://www.ncbi.nlm.nih.gov/pmc/articles/PMC4069282/.

200 **interregional connectivity strength predicts interregional activity correlations:** https://www.biorxiv.org/content/10.1101/422477v2.

200 **anhedonia for music in human beings:** https://www.ncbi.nlm.nih.gov/pmc/
articles/PMC5135354/.

200 **brain region representing hierarchies of kinship relationships:** https://www
.nature.com/articles/s41467-020-16489-x/.

200 **gene expression patterns that determine how cellular diversity and axon
guidance implement brain wiring:** https://www.ncbi.nlm.nih.gov/pmc/
articles/PMC6086934/; https://www.ncbi.nlm.nih.gov/pmc/articles/
PMC6447408/; https://www.biorxiv.org/content/10.1101/2020.03.31.016972v2;
https://www.biorxiv.org/content/10.1101/2020.07.02.184051v1; https://www.ncbi
.nlm.nih.gov/pmc/articles/PMC5292032/.

203 **the *Moro* reflex:** https://en.wikipedia.org/wiki/Moro_reflex.

EPILOGUE

205 **the very same brainstem tumor that had grown in the little girl with eyes
misaligned:** https://www.ncbi.nlm.nih.gov/pmc/articles/PMC5891832; https://
www.ncbi.nlm.nih.gov/pmc/articles/PMC5462626; https://www.ncbi.nlm.nih
.gov/pmc/articles/PMC6214371.

206 **the last Neanderthal may have been also the last survivor of a wave of modern humans:** https://www.ncbi.nlm.nih.gov/pmc/articles/PMC4933530/;
https://www.biorxiv.org/content/10.1101/687368v1.

207 **exploring how the light-gated protein called channelrhodopsin actually
works:** https://www.ncbi.nlm.nih.gov/pmc/articles/PMC5723383/; https://
www.ncbi.nlm.nih.gov/pmc/articles/PMC6340299/; https://www.ncbi.nlm
.nih.gov/pmc/articles/PMC6317992/; https://www.ncbi.nlm.nih.gov/pmc/
articles/PMC4160518/.

208 **research into microbial light responses over the past 150 years:** https://www
.ncbi.nlm.nih.gov/pmc/articles/PMC5723383/.

209 **like the performance artist Joan Jonas said about her art in 2018:** https://
twitter.com/KyotoPrize/status/1064378354168606721.

210 **depending on the study or population, from 1 to 7 percent:** https://www
.ncbi.nlm.nih.gov/books/NBK55333/.

211 **An easy explanation is that extinction follows very quickly from technology:** https://en.wikipedia.org/wiki/Fermi_paradox.

211 **There is a heritable component:** https://www.ncbi.nlm.nih.gov/pmc/articles/
PMC6309228/; https://www.ncbi.nlm.nih.gov/pmc/articles/PMC5048197/.

211 **linked to sociopathy or aggression:** https://www.ncbi.nlm.nih.gov/pmc/
articles/PMC2430409/; https://www.ncbi.nlm.nih.gov/pmc/articles/
PMC6274606/; https://www.ncbi.nlm.nih.gov/pmc/articles/PMC6433972/;
https://www.ncbi.nlm.nih.gov/pmc/articles/PMC5796650/.

212 **a frenzy of violent aggression toward another mouse:** https://www.ncbi.nlm
.nih.gov/pmc/articles/PMC3075820/.

214 the fascinating domain of the philosophical treatise: https://www.science direct.com/science/article/pii/S0896627313011355?via%3Dihub.

214 In 2020, recording the activity of cells broadly across mouse and human brains: https://www.ncbi.nlm.nih.gov/pmc/articles/PMC7553818/.

215 control of cells by type—the workhorse of optogenetics for its first fifteen years: https://www.ncbi.nlm.nih.gov/pmc/articles/PMC5296409/.

215 but also allowing control of activity in many single cells, or individually specified neurons: https://www.ncbi.nlm.nih.gov/pmc/articles/PMC5734860/; https://www.ncbi.nlm.nih.gov/pmc/articles/PMC3518588/.

215 Now we can pick, at will, tens or hundreds of single cells for optogenetic control: https://www.ncbi.nlm.nih.gov/pmc/articles/PMC6447429/; https://www.ncbi.nlm.nih.gov/pmc/articles/PMC6711485; https://www.biorxiv.org/content/10.1101/394999v1.

215 we can pick the cells that normally respond to vertical (but not horizontal) stripes: https://www.ncbi.nlm.nih.gov/pmc/articles/PMC6711485.

217 new life has even been breathed into philosophical thought experiments: https://en.wikipedia.org/wiki/Einstein%27s_thought_experiments.

PERMISSIONS

KARL DEISSEROTH is professor of bioengineering and psychiatry at Stanford University. He received his undergraduate degree summa cum laude at Harvard, and his MD/PhD at Stanford, where he completed psychiatry training and is board-certified by the American Board of Psychiatry and Neurology. Deisseroth is known for creating and applying new technologies for studying the brain, including optogenetics—for which he was the winner of the 2018 Kyoto Prize and the 2020 Heineken Prize, among many other major international awards. Deisseroth has five children and lives near Stanford University, where he teaches and directs Stanford's undergraduate degree in bioengineering, and treats patients with mood disorders and autism. Deisseroth helped craft the multibillion-dollar ongoing U.S. national BRAIN Initiative, and is a member of the National Academy of Medicine, the National Academy of Sciences, and the National Academy of Engineering.

ABOUT THE TYPE

This book was set in Electra, a typeface designed for Linotype by W. A. Dwiggins, the renowned type designer (1880–1956). Electra is a fluid typeface, avoiding the contrasts of thick and thin strokes that are prevalent in most modern typefaces.